HUMANOID ROBOTICS and NEUROSCIENCE
SCIENCE, ENGINEERING and SOCIETY

FRONTIERS IN NEUROENGINEERING

Series Editor
Sidney A. Simon, Ph.D.

Published Titles

Humanoid Robotics and Neuroscience: Science, Engineering, and Society
Gordon Cheng, ATR Computational Neuroscience Labs, Kyoto, Japan

Methods in Brain Connectivity Inference through Multivariate Time Series Analysis
Koichi Sameshima, University of São Paulo, São Paulo, Brazil
Luiz Antonio Baccala, University São Paulo, São Paulo, Brazil

Neuromorphic Olfaction
Krishna C. Persaud, The University of Manchester, Manchester, UK
Santiago Marco, University of Barcelona, Barcelona, Spain
Agustín Gutiérrez-Gálvez, University of Barcelona, Barcelona, Spain

Indwelling Neural Implants: Strategies for Contending with the *In Vivo* Environment
William M. Reichert, Ph.D., Duke University, Durham, North Carolina

Electrochemical Methods for Neuroscience
Adrian C. Michael, University of Pittsburg, Pennsylvania
Laura Borland, Booz Allen Hamilton, Inc., Joppa, Maryland

HUMANOID ROBOTICS and NEUROSCIENCE
SCIENCE, ENGINEERING and SOCIETY

Edited by

Gordon Cheng, PhD

CRC Press
Taylor & Francis Group
Boca Raton London New York

CRC Press is an imprint of the
Taylor & Francis Group, an **informa** business

CRC Press
Taylor & Francis Group
6000 Broken Sound Parkway NW, Suite 300
Boca Raton, FL 33487-2742

First issued in paperback 2019

© 2015 by Taylor & Francis Group, LLC
CRC Press is an imprint of Taylor & Francis Group, an Informa business

No claim to original U.S. Government works

ISBN-13: 978-1-4200-9366-7 (hbk)
ISBN-13: 978-0-367-37789-2 (pbk)

Visit the Taylor & Francis Web site at
http://www.taylorandfrancis.com

and the CRC Press Web site at
http://www.crcpress.com

Contents

SECTION I   Humanoid Robotics Perspective to Neuroscience

SECTION II   Emulating the Neuro-Mechanisms with Humanoid Robots

SECTION III   Leaping Forward: Toward Cognitive Humanoid Robots

Series Preface

The *Frontiers in Neuroengineering* series presents the insights of experts on emerging experimental techniques and theoretical concepts that are or will be at the vanguard of neuroscience. Books in the series cover topics ranging from electrode design methods for neural ensemble recordings in behaving animals to biological sensors. The series also covers new and exciting multidisciplinary areas of brain research, such as computational neuroscience and neuroengineering, and describes breakthroughs in biomedical engineering. The goal is for this series to be the reference that every neuroscientist uses to become acquainted with new advances in brain research.

Each book is edited by an expert and consists of chapters written by leaders in a particular field. The books are richly illustrated and contain comprehensive bibliographies. The chapters provide substantial background material relevant to the particular subject.

We hope that, as the volumes become available, our efforts as well as those of the publisher, the book editors, and the individual authors will contribute to the further development of brain research. The extent to which we achieve this goal will be determined by the utility of these books.

Sidney A. Simon, Ph.D
Series Editor

The Editor

Gordon Cheng holds the Chair of Cognitive Systems, Founder and Director of Institute for Cognitive Systems, at the Technical University of Munich. From 2002–2008, he was the Head of the Department of Humanoid Robotics and Computational Neuroscience, ATR Computational Neuroscience Laboratories, Kyoto, Japan. He was the Group Leader for the newly initiated JST International Cooperative Research Project (ICORP), Computational Brain. He has also been designated as a Project Leader/Research Expert for the National Institute of Information and Communications Technology (NICT) of Japan. He is also involved (as an adviser and as an associated partner) in a number of major European Union projects.

He held fellowships from the Center of Excellence (COE), Science and Technology Agency (STA) of Japan (1998–2002). Both of these fellowships were taken at the Humanoid Interaction Laboratory, Intelligent Systems Division at the Electro Technical Laboratory (ETL), Japan. At ETL he played a major role in developing a completely integrated humanoid robotics system. He received his Ph.D. in systems engineering (in 2001) from the Department of Systems Engineering, The Australian National University, and his Bachelor and Master degrees in computer science (respectively in 1991 and 1993) from the University of Wollongong, Australia. He has extensive industrial experience in consultancy as well as contractual development of large software systems. From 1994–2001, he was also the Director of G.T.I. Computing, a company he founded, which specializes in networking and transport management systems in Australia.

His research interests include humanoid robotics, cognitive systems, neuroengineering, real-time network robot control, brain–machine interfaces, biomimetics of human vision, computational neuroscience of vision, action understanding, human–robot interaction, active vision, mobile robot navigation, and object-oriented software construction.

Contributors

Gordon Cheng, Ph.D.
Institute for Cognitive Systems
Technische Universität München
München, Germany

Sang-Ho Hyon, Ph.D.
Department of Robotics
College of Science and Technology
Ritsumeikan University
Kyoto, Japan

Jun Morimoto, Ph.D.
Department of Brain Robot
    Interface(BRI)
ATR Computational Neuroscience
    Laboratories
Kyoto, Japan

Erhan Öztop, Ph.D.
Computer Science Department
Ozyegin University
Istanbul, Turkey
and
Advanced Telecommunication
Research Institute International
Kyoto, Japan
and
Biological ICT group
National Institute of Information and
    Communications Technology
Kyoto, Japan

Tetsunari Inamura, Ph.D.
The Graduate University for
    Advanced Studies
National Institute of Informatics
Tokyo, Japan

Michael Mistry, Ph.D.
School of Computer Science
University of Birmingham
Birmingham, United Kingdom

Aleš Ude, Ph.D.
Laboratory of Humanoid and
    Cognitive Robotics
Jožef Stefan Institute
Ljubljana, Slovenia

Stefan Schaal, Ph.D.
Computational Learning and Motor
    Control Lab
University of Southern California
Los Angeles, USA

Helge Ritter, Ph.D.
Faculty of Technology and Excellence
    Cluster Cognitive Interaction
    Technology (CITEC)
Bielefeld University
Bielefeld, Germany

Robert Haschke, Ph.D.
Faculty of Technology and Excellence
    Cluster Cognitive Interaction
    Technology (CITEC)
Bielefeld University
Bielefeld, Germany

Yasuo Kuniyoshi, Ph.D.
Department of Mechano-Informatics
School of Information Science
    and Technology
The University of Tokyo
Tokyo, Japan

Minoru Asada, Ph.D.
Department of Adaptive
    Machine Systems
Graduate School of Engineering
Osaka University
Osaka, Japan

Yoshihiko Nakamura, Ph.D.
Department of Mechno-Informatics
School of Information Science
  and Technology
The University of Tokyo
Tokyo, Japan

Emre Ugur, Ph.D.
Institute of Computer Science
University of Innsbruck
Innsbruck, Austria
and
Advanced Telecommunication
  Research Institute International
Kyoto, Japan
and
Biological ICT group
National Institute of Information and
  Communications Technology
Kyoto, Japan

Yu Shimizu, Ph.D.
Biological ICT group
National Institute of Information and
  Communications Technology
Kyoto, Japan
and
Neural Computation Unit
Okinawa Institute of Science
  and Technology
Graduate University
Okinawa, Japan

Hiroshi Imamizu, Ph.D.
Department of Cognitive Neuroscience
ATR Cognitive Mechanisms
  Laboratories
Kyoto, Japan
and
Center for Information and
  Neural Networks
National Institute of Information and
  Communications Technology
Kyoto, Japan

Section I

Humanoid Robotics Perspective to Neuroscience

1 Humanoid Robotics and Neuroscience: Science, Engineering, and Society

Gordon Cheng

CONTENTS

1.1   INTRODUCTION

Ever since the dawn of civilization, we as humans have been fascinated with machines and devices that can replicate aspects of biology, in particular of ourselves. Some are created for our entertainment, some to facilitate us in our daily lives, and historically speaking some were even created for imitating the power of gods (religious relics) [1]. The themes of these developments have gone in and out of trends in various forms, but the most fundamental issues were to explore points toward the eventuation of robotics as we know it today.

We are on the verge of a new era of rapid transformations in both science and engineering, a transformation that brought together technological advancements in a fusion that shall accelerate both **science** and **engineering**. This new transformation brings together scientists working under a new direction of robotic research.

The utility of robots holds great promise not only in industrial automation; more recently it has also been taken on by neuroscientists as a tool to aid in the discovery of mechanisms in the human brain. In particular, with the emergence of numerous advanced humanoid robots, unlike usual robotic systems, these are highly sophisticated humanlike machines equipped with humanlike sensory and motor capabilities. These robots are now among us, contributing to our scientific endeavors.

Aiming at better assisting mankind has motivated engineers to look more closely at other scientific findings for the creation of innovative solutions that could better co-exist in our common **society**.

1.1.1   OUR RESEARCH PARADIGM

In essence, here we advocate three essential interconnecting themes in our research paradigm:

- In *science:* Building a humanlike machine and the reproduction of humanlike behaviors can in turn teach us more about how humans deal with the world, and the plausible mechanisms involved.
- In *engineering:* Engineers can gain a great deal of understanding through the studies of biological systems, which can provide guiding principles for developing sophisticated and robust artificial systems.
- For *society:* In so doing we will gain genuine knowledge toward the development of systems that can better serve our society.

Such an approach will examine in depth various issues that go beyond the pure engineering of a robot and will require the integration of multiple disciplines in exploring and exploiting what has been learned from other fields such as philosophy, neuroscience, psychology, and physiology among others. It is foreseeable that this multidisciplinary integration approach will examine in depth:

- How do humans handle all different types of interaction with ease and in such a competent manner?

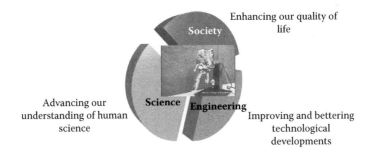

FIGURE 1.1   Research paradigm: Science, engineering, and society. See color insert.

- How can such a rich system be built?
- What are the underlying mechanisms?
  - What are the underlying processes and controls?
- How can we benefit from this approach?

1.1.2 ROBOTS IN THE REAL WORLD

Recently robots have been moving off the factory floors and into our homes. Possibly one day these robotic systems will help us in our daily lives, also as ideal research tools. During the late 1990s a revitalization of interest in the building of humanlike robots has emerged. This resurgence of interest has produced a vast number of spectacular humanoid systems. Here we highlight some of these recent systems.

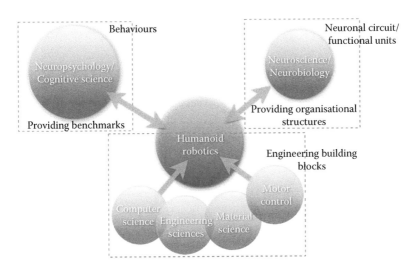

FIGURE 1.2   A multidisciplinary integration approach bringing together robotics and neuroscience.

FIGURE 1.3   SONY robotic dog, AIBO. (Courtesy of SONY. With permission.)

1.1.2.1   Entertainment Robots

Electronic SONY was one of the biggest manufacturers who introduced one of the first devices for home robot entertainment. In 1999, SONY introduced a four-legged doglike robot, AIBO, into the mass market as companions for people. These small-size entertainment (see Figure 1.3) robots have enjoyed worldwide acceptance. In the 2000s, SONY introduced a humanlike robot as the next generation of entertainment robots to the world (see Figure 1.4). It was a highly sophisticated integrated system, equipped with stereo cameras for eyes, microphones for ears, and an array of capabilities including walking and dancing [2].

Animatronics, one of the first nonindustrial companies in the area of animatronics, and entertainment companies such as Disney and MGM Studios are just a few who

FIGURE 1.4   SONY humanoid robot SDR-4. (Courtesy of SONY. With permission.)

FIGURE 1.5   Iguana, an animatronics entertainment robot. (Courtesy of Stephen C. Jacobsen, SARCOS. With permission.). An animatronics character in the form of an iguana, operates on a daily basis at the RC Willey Restaurant.

have utilized robotic technology within part of the venue to entertain all ages. One active group that has been supplying sophisticated robotic systems is the Utah company SARCOS. Their range of development includes full-body animatronics figures that can produce realistic human motions. They have also developed robotic animals from insects, singing birds, and iguanas (see, e.g., Figure 1.5) to full-size human figures.

A SARCOS humanoid figure (see Figure 1.6) [3] was designed specifically for the FORD Motor Company in 1995; this highly sophisticated system traveled to motor shows across North America and Europe from 1995 to 1997.

1.1.3 HUMANOID ROBOTS IN RESEARCH

In this section we present some existing humanoid systems in research and briefly outline some of their approaches and achievements.

In December 1996, one of the highest impacts to robotics in recent times was made by the Japanese motor company Honda Motor Co., Ltd.: the release and announcement of a 15-year project which produced a full-size humanoid robotic system (see Figure 1.7) that was able to walk autonomously and climb stairs [4, 5]. This level of achievement set a standard for the engineering of highly sophisticated humanoid robots for years to follow.

1.1.3.1 Humanoid Service Robots

One avenue of humanoid research has been considering humanoids as ideal service robots. Taking the view that much of our everyday environment has been specifically designed for the use of humans, these humanlike robots would ideally be suitable for performing daily chores usually done by humans. This type of robot has been considered to be the most suitable to assist mankind with daily activities.

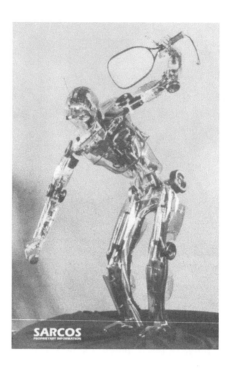

FIGURE 1.6   The FORD. (Courtesy of Stephen C. Jacobsen, SARCOS. With permission.)

The humanoid robot ARMAR (see Figure 1.8a) is a series of humanoid robots developed at the Karlsruhe Institute of Technology (KIT) [6, 7]. It was developed in targeting the introduction of such robots in the kitchen environment; for instance, it is even capable of loading dishes in a dishwasher.

The humanoid robot HERMES (see Figure 1.8b), developed by the Institute of Measurement Science, Bundeswehr University Munich, Germany [8, 9] provided another interesting aspect of service robots; one of the primary aspects of this particular system is that they aim for reliability. This system is able to text-to-speech interface. The AMI humanoid robot from the Korea Advanced Institute of Science and Technology (KAIST) [10] is a humanoid that is capable of a few household tasks, such as vacuuming (see Figure 1.8c).

At a further extreme both the Japanese and the Korean governments have decided to support projects related to humanoid robots that can facilitate society. A five–year project supported by the Ministry of Economy, Trade and Industry (METI) through the New Energy and Industrial Technology Development Organization (NEDO) of Japan, the Humanoid Robotics Project (HRP) was established to investigate possible "applications" for humanoid robots. They set out to examine five scenarios: (1) maintenance tasks of industrial plants, (2) teledriving of construction machines, (3) security service at home and office, (4) taking care of patients, and (5) cooperative works in the open air [11, 12]. The total robotic system was designed and integrated by Kawada Industries, Inc. together with the Humanoid Research Group of the National Institute of Advanced Industrial Science and Technology (AIST) [13].

FIGURE 1.7   Humanoid robots P3 and ASIMO (Advance). (Courtesy of HONDA. With permission.)

The NASA Johnson Space Center in the United States is developing a humanoid upper body robot for use in space. They have designed the humanoid robot, Robonaut [14]. The primary objective of this system is to perform extravehicular activity (EVA). This sets the scene for future service robots in space.

1.1.3.2   Humanoid Robots as Research Tools

Although most robotic systems presented here are from research laboratories, many of them aim not just to build humanoid robots for the sake of constructing better machines. A number of these research laboratories are investigating issues beyond

FIGURE 1.8   Service-style humanoid robots. (a): ARMAR. (Courtesy of Tamim Asfour. With permission.); (b) HERMES. (Courtesy of Rainer Bischoff. With permission.); (c) AMI humanoid robot. (Courtesy of Hyun S. Yang, KAIST. With permission.)

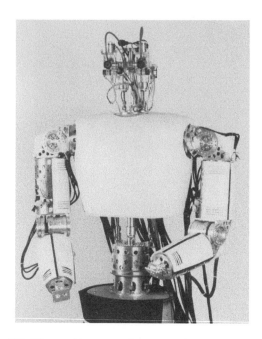

FIGURE 1.9   The ETL-Humanoid system. (Courtesy of Yasuo Kuniyoshi. With permission.) See color insert.

the engineering of systems; exploration is also underway in areas such as intelligent/cognition as well as basic science (thus, the research paradigm of "understanding through creating").

1.1.3.3   Cognition and Social Interaction

The MIT project, COG [15], was one of the first and noticeable groups that proposed to set out with the ambition of creating a humanoid robot that exhibits various aspects of "cognition" [16], ranging from basic visual and auditory attention [17], through various aspects of childlike development [18]. Following a similar research approach, in the work of Kawamura et al., they have been gradually building a humanoid system that has similar cognitive ability to that of humans; their main goal is to develop a system that can support elderly patients [19].

Around the end of the 2000s, a multimodal interactive humanoid system was developed as a platform for the study of humanoid interaction [20]. The system, ETL-Humanoid (see Figure 1.9), was developed at the ElectroTechnical Laboratory (ETL) in Japan during the period of 1996–2001 [21]. This system was designed as a research tool for exploring general principles of intelligent systems through interaction with the changing world. Interactions involving visual, auditory, and the physical all form part of its integration; one noticeable work yielded complex and meaningful interaction with a human through exploiting a continuous sensory–motor cooperative–competitive integration architecture, which demonstrated continuous adaptivity, redundancy, and flexibility [22].

FIGURE 1.10   Babybot from the LiraLab. (Courtesy of Giorgio Metta. With permission.)

At the same time, the LiraLab at the University of Genoa, Italy was attempting to combine ideas from human development in the construction of their humanlike system, Babybot (see Figure 1.10) [23]. Their robot learns through stages of development to perform the basic task of reaching. One especially interesting feature of this system is that the vision processing is performed through the use of a log-polarlike CMOS camera, having pixels arranged in a configuration emulating that of the human retina, from the center to the periphery in dense to sparse pixel arrangements. This emulation shows that computation and information flow can be reduced, providing the same visual capabilities as these of a child. Their approach makes a detailed attempt in providing and demonstrating benefits of emulating details of biology in the development of a sophisticated human-like robot [24].

A noticeable followup to the Babybot project is the European project RoboCub. This integrated project aimed to develop a 3.5-year–old child-sized humanoid robot, the iCub (see Figure 1.11), for the study of cognition through its implementation of cognitive capabilities similar to those of a child [25]. Unlike many past projects, this 5-year–long project continued to provide iCub as a humanoid robotic platform for research; more than 20 iCub robots have been deployed in facilitating research worldwide.

1.1.3.4   Social Interaction

One noticeable work in the area of social interaction is by a group in Japan. Investigating the use of robots to interact with autistic children, their aim or wish is ultimately to draw out children with identifiable social defects and teach them to interact in more humanlike ways. The robot is called "Infanoid" (see Figure 1.12) [26] and was developed as a tool for this purpose. Preverbal communication in infants, that is, using nonverbal means (e.g., gaze, gestures, etc.), plays a crucial role in human communication development. Children with autism who show typical communication disorders cannot use these preverbal communication skills, which leads to serious impairment

FIGURE 1.11   The iCub humanoid robot (iCub at the TUM-ICS lab). See color insert.

FIGURE 1.12   Infanoid. (Courtesy of Hideki Kozima. With permission.)

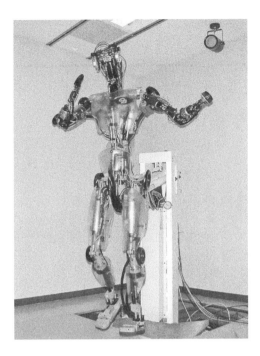

FIGURE 1.13   ATR humanoid robot, DB (co-developed with SARCOS during the JST Kawato Dynamic Brain Project). (Courtesy of Stefan Schaal. With permission.)

in social abilities and verbal communication. The aim of this ongoing project is to understand the mechanism/development of human communication.

1.1.3.5 Humanoid Robot in Neuroscience

The Advance Telecommunication Research Institute was the first to propose that a humanoid robot could actually contribute to neuroscience studies. In their work they demonstrated several aspects of humanlike learning, and were successful in applying these to humanoid robots [27]. The humanoid robot, DB (dynamic brain), was specifically developed for this purpose (see Figure 1.13). The success of this work set an important landmark in the application of scientific interchange between engineers and neuroscientists.

1.1.4 CB—Computational Brain

Following the guiding principles above, a 50-degrees-of-freedom humanoid robot, CB, computational brain, was realized. CB is a humanoid robot created for exploring the underlying processing of the human brain while dealing with the real world. We place our robotic investigations within real-world contexts, as humans do. In so doing, we focus on utilizing a system that is closer to humans in sensing, kinematics configuration, and performance.

FIGURE 1.14   The humanoid robot CB (computational brain).

We present the real-time network-based architecture for the control of all 50 degrees-of-freedom. The controller provides full position/velocity/force sensing and control at 1 KHz, allowing us the flexibility in deriving various forms of control. A dynamic simulator is also presented: the simulator acts as a realistic testbed for our controllers, and acts as a common interface to our humanoid robots. A contact model developed to allow better validation of our controllers prior to final testing on the physical robot is also presented.

Three aspects of the system are highlighted in this chapter (1) physical power for walking; (2) full-body compliant control, physical interactions; and (3) perception and control, visual ocular–motor responses.

Our objective is to produce a richly integrated platform for the investigation of humanlike information processing, exploring the underlying mechanisms of the human brain in dealing with the real world. In this chapter, we present a humanoid robotic system, a platform created to facilitate our studies.

Our focus is toward the understanding of humans, more specifically the human brain, and its underlying mechanisms in dealing with the world. We believe that a humanoid robot that is closer to a human being will facilitate this investigation. Such a sophisticated system will impose the appropriate constraints by placing our exploration within the context of human interactions and human environments. As a result, a full-size humanoid robot CB (computational brain) was built to match closely the physical capability of a human, thus making it suitable for the production of a variety of humanlike behaviors, utilizing algorithms that originate in computational neuroscience.

1.1.4.1   Outline

The following sections describe the physical robotic system and the supporting software control architecture used in our research. We present experimentally three

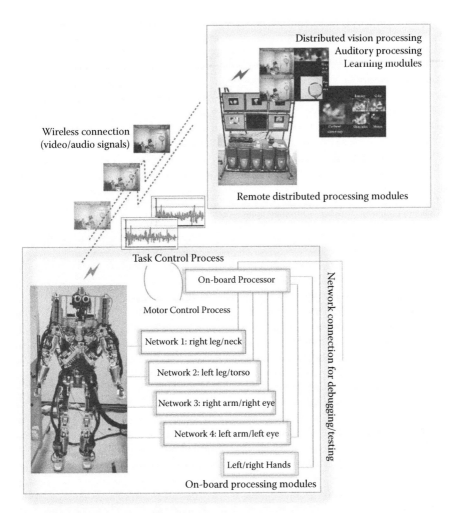

FIGURE 1.15   Overview of the CB research platform and setup, providing full support for local processing for robot sensing and motor control, also showing that demanding high-level processes are dealt with using remote distributed processors.

aspects of our system: (1) adequate performance; (2) force controllability; and (3) perceptual abilities of our humanoid system.

1.1.5   Research Platform—Hardware and Software Architecture

In this section, a presentation of the hardware and software architecture of our research platform, CB, is presented. An overview of the setup is depicted in Figure 1.15, Table 1.1 which, presents the overall technical specifications of the system and explains their corresponding biological counterparts.

TABLE 1.1
Overall Specification of Humanoid Robot—CB

Degrees of Freedom	50 in Total	
Actively compliant	Arms/Legs/Torso/Neck (34 DOFs)	Central body parts of the humanoid robot
Passively compliant	Fingers/Eyes (16 DOFs)	Forming two hands for manipulation
Weight	92Kg	
Height	157.5cm	Average person
Orientation sensors	2 × 6 DOF (translational and rotational)	Emulating the vestibular system
Foot force sensors	2 × 6 DOF (left and right)	Ground contact detection
Onboard computer	Arbor PC-104 plus Em104P-i7013/PM1400 1.4 GHz Intel Pentium-M Processor	Acting as the motor controller of the whole system, emulating the role of the spinal column
Cameras	2× ELMO MN42H 17mm O.D. (Peripheral)/2× ELMO QN42H 7mm O.D. (Foveal)	Emulating both the peripheral and foveal human visual systems
Microphones	2× SHURE Model MX180	Emulating human binaural hearing system

1.1.5.1   Humanoid Robot—CB

The humanoid robot CB was designed with the general aim of developing a system capable of achieving human capabilities, especially in its physical performance. CB, the physical system, is of general human form; the following sections present the basics of the system.

1.1.5.1.1   Mechanical Configuration

CB is a full-body humanoid robot. It is approximately 157.5 cm in height and approximately 92 kg in weight. It has an active head system with 7 degrees of freedom (2 × 2 degrees-of-freedom eyes, 1 × 3 degrees-of-freedom neck), 2 × 7 degrees-of-freedom arms, 2 × 7 degrees-of-freedom legs, 1 × 3 degrees-of-freedom torso, and 2 × 6 degrees-of-freedom hands (see Figure 1.3), 50 degrees of freedom in total (see Figure 1.1). The system has similar ranges of motion and physical performance as a human person (as guided by human factors studies [19]). The system is able to perform saccadic eye movements at up to 3 Hz (similar to that of humans). The hands (as shown Figure 1.3) have been developed to provide basic functionality such as grasping, pointing, and pinching.

1.1.5.2   Sensing Subsystems

The active head houses a set of inertial sensors (three-axis rotational gyro, three-axis translational accelerometer). They are used to emulate the human vestibular system (the inner ear), providing head orientation, as used for gaze-stabilization. An

SARCOS Private and Proprietary

FIGURE 1.16   Five-finger humanoid robot hand with six degrees of freedom (built by SAR-COS).

additional inertial sensor is installed at the hip to provide angular velocity/translational acceleration near the center of mass of the whole body, used to provide the overall orientation of the system. The visual system is made up of two cameras per eye; peripheral (wide-angle view) and foveal (narrow view) cameras are installed on each eye to emulate the visual acuity of the human visual system. Stereo microphones have also been installed to provide hearing to the robot. Images from the video cameras and the auditory signals are transmitted over high-speed wireless communication to a network of PCs for perceptual processing.

1.1.5.3   Proprioceptive Sensing

Proprioceptual information plays a key role in human motor control, informing the limbs and higher-level cortices of critical information in carrying out suitable action. Our system is equipped to support various forms of control to account for interactions; position/velocity/torque sensing are provided at the key joints for proper active compliant control (arms/legs/torso/neck, 34 DOFs). Foot force sensors are installed at the soles of each foot to provide information during ground contact and weight distribution, as it is critical for walking and balancing control. The uniqueness of this system over previous systems is that torque sensors are installed on the main joints of the system, the arms/legs, torso and the neck, allowing joint-level active compliance possible.

1.2   SOFTWARE CONTROL ARCHITECTURE

As depicted in Figure 1.17, we developed a network-based architecture to better enable us to explore various levels of humanlike processing on our humanoid system. It is divided into two: onboard low-level computing and higher-level perceptual processing.

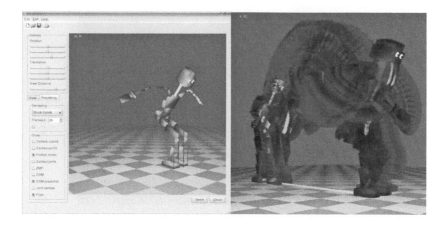

FIGURE 1.17   Snapshot of our dynamic simulator.

1.2.1 ONBOARD COMPUTING

The control of the robot is through an onboard PC104-plus CPU stack with an Intel 1.4 GHz Pentium-M processor, which provides control and sensory feedback to all the joints, performed at a maximum rate of 1Khz (for all 50 degrees of freedom). The motor control process 2 gathers sensory data (joint position/velocity/force information, foot force sensors, gyro/accelerometer data) of the four internal network branches to each of the low-level joint controllers (as depicted in Figure 1.17). A task-level control process 3 running on the local processor communicates with the motor control process to perform higher-level control functions. One external 100 Mbits/s Ethernet network connection is also provided for debugging and testing purposes. The onboard computer is running the ART-Linux operating system, a real-time version of Linux, which retains all the standard functionality of Linux with the addition of real-time support [28]. The onboard processor is sufficient for walking control and force-based balancing control. Any processing that requires substantial computation such as vision processing is performed remotely on a cluster of PCs (see next section).

1.2.1.1 Dynamic Simulation and Ubiquitous Interface

Having a testbed is an essential element in conducting research, and for validating a variety of control strategies and algorithms for a complex physical system such as a humanoid robot. A realistic dynamic simulator has been developed for our humanoid robot [29]. The simulator serves two purposes: realistic dynamic simulation, providing a realistic testbed for our controllers prior to experimentation on the real robot; and an ubiquitous interface, providing a transparent programming interface between the simulator and the real robot.

TABLE 1.2
Accessible Data Within Simulator

Joint position	Arms/Legs/Torso/Neck/Eyes (38 DOFs)
Joint torque	Arms/Legs/Torso/Neck (34 DOFs)
Joint position, velocity, and torque control gains	Arms/Legs/Torso/Neck (34 DOFs)
Foot force contact	2×6 DOFs (left/right)
Contact points	8×3 DOFs
COM (center of mass)	3 DOFs
Zero moment point	3 DOFs
Friction parameters	2 DOFs (static and dynamic)
Link coordinate frames	38 Homogeneous transform matrices
COM (center of mass) Jacobian	DOFs $\times$ 3 matrix
Contact point Jacobians	8 DOFs $\times$ 3 matrices

1.2.1.2 A Realistic Dynamic Simulator

Our dynamic simulator serves as a realistic testbed for our controllers prior to experimentation on the real robot. One key aspect of the simulator is the development of a new contact model [29]. The contact model was developed to facilitate better evaluation of our controllers in a safe and consistent software module manner by quickly computing accurate contact interactions based on empirical estimates of real-world friction properties. Figure 1.17 shows our CB simulator.

The simulator provides kinematic and dynamic analyses for task control. It provides access to the robot's posture, link coordinate frames, dynamic Jacobians, ZMP, and so on, based on sensor data (see Table 1.2).

Modeling contact resolution methods have been extensively investigated to support our simulator [29]. In our CB simulator, an efficient contact handling method that applies Coulomb friction exactly has been developed. It is important for humanoid robots to handle contact and friction forces accurately and efficiently in order to validate their controllers for real-world conditions. The method used in the CB simulator models contact between the humanoid's feet and a planar ground surface, and explicitly identifies the contact state of each foot. For a rectangular-footed biped, there are 361 possible contact states. Each foot may be contacting in one of nine ways: at one of four vertices, one of four edges, with its face, or not in contact at all. Active contacts may be static or dynamic.

Friction parameters may be specified during simulation. Static friction is applied by incorporating the appropriate motion constraints into forward dynamics computations using Lagrange multipliers, and dynamic friction is integrated as additional forces. Measurements of the friction properties of CB's feet with various surfaces were conducted in order to provide an empirical basis for accurate simulation of the robot's contact interactions with its environment.

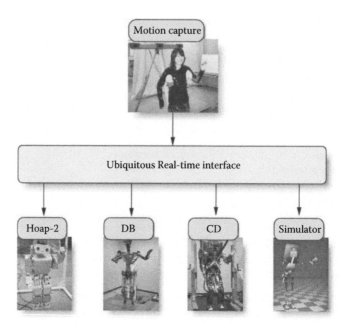

FIGURE 1.18   General concept of our ubiquitous interface.

1.2.1.3   Ubiquitous Interface

The simulator also acts as a common interface to our collection of humanoid robots. Controller code running a simulated robot can be executed identically on the corresponding real robot. This accords with the concept of a humanoid robotics platform described in Reference [13]; that is, it provides a seamless control interface between the simulated and actual robot. Also, the provision of an ubiquitous control environment ideally facilitates an identical software interface to arbitrary robots. This ensures the portability and generalized utility of control software. In our system we provide this functionality by using an identical network interface for communication between the motor control process and the simulator. Figure 1.18 illustrates the general idea of our ubiquitous interface.

1.2.1.4   Architecture for Emulating Humanlike Processing

The overall structure of the system has been developed to facilitate the emulation of human information processing. The distributed architecture we proposed is motivated by neurological studies of the visual cortices (e.g., Van Essen et al. [30]). These findings showed in great detail that processing in the brain is a well-organized structure with connections and pathways. In our system, we emulated this aspect of the organizational structure with a network of computers. The cluster of PCs is connected via a series of 1 Gbits/s Ethernet networks. We proposed a framework utilizing the UDP protocol, data streamed from one process to another. The processing on each PC can range from the most basic (e.g., color extraction, edge filtering, etc.) to higher levels (e.g., visual tracking, recognition, etc.). The sophistication of the system can increase quite rapidly simply through connecting the processing outputs of simpler elements to the inputs of more advanced processing elements in a bottom-up manner. Manipulation of the lower-level processes can also be performed in a top-down fashion. This framework will provide the capability to explore a greater range of cognitive architectures. Modules from spatial hearing to learning modules (SOM, reinforcement learning, associative memory) have also been developed to support our studies.

The effectiveness of this distributed architecture has been demonstrated with the development of a visual attention system [31], a real-time human hand tracking system [32], and visual ocular–motor control for our humanoid system.

1.2.2 HUMANLIKE BEHAVIORS

In this section, we present experimentally three aspects of our system: (1) physical power for walking; (2) full-body compliant control, physical interactions; and (3) perception and control, visual processing and ocular–motor responses.

1.2.2.1 Humanlike Performance—Walking

To examine humanlike walking, we have investigated a number of biologically inspired walking algorithms [33–35]. Here we present the central pattern generator (CPG) based locomotive controller that managed to produce natural-looking walking movements on our humanoid robot.

In the experiment shown in Figure 1.19, we demonstrated that a humanoid robot could step and walk using simple sinusoidal desired joint trajectories with their phase adjusted by a coupled oscillator. The centers of pressure and velocity are used to detect the phase of the lateral robot dynamics. This phase information is then utilized to modulate the desired joint trajectories, thus enabling us to generate successful stepping and walking patterns [36]. This walking demonstration shows that our system has adequate performance in supporting the full body.

1.2.2.2 Compliant Control: 3D Balancing and Physical Interactions

Force control or compliant control plays a key role in the ways humans interact with the world. This is well supported by biomechanical studies [37]. Humans control the stiffness of their limbs in dealing with contacts while performing appropriate tasks. This is also supported by the passivity of nonlinear control theory [38]. We developed a controller to perform force-based control of CB in 3D.

One of the controllers we developed is a full-body contact force controller with gravity compensation [39]. The gravity compensation is possible only for a force-controllable humanoid like CB; an example of the force tracking performance of the system is shown in Figure 1.21. This makes the robot passive with respect to external force applied to arbitrary contact points; hence, it results in compliant full-body interaction. The additional contact force allows the robot to generate the desired ground reaction forces and other necessary interaction forces. For example, if we define the desired ground reaction force as a feedback to the center of mass (CoM), then full-body 3D balancing is easily realized. Figure 1.20 shows such an

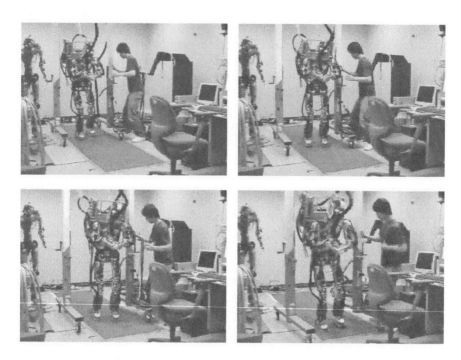

FIGURE 1.19   Walking like a human, utilizing a biological-based CPG controller.

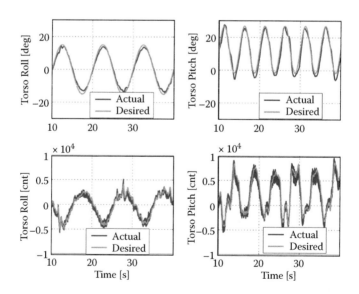

FIGURE 1.20   Force-based tracking performance of our force-based controller during 3D balancing. Upper graphs show the angle commands and the responses of the torso joints ("figure-eight" trajectory) while maintaining balance. Lower graphs show the corresponding force inputs and the "Actual" force feedback.

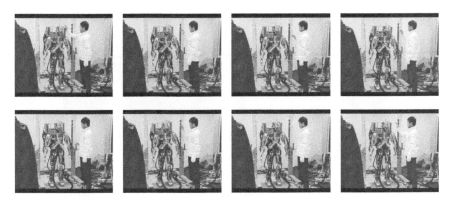

FIGURE 1.21   Physical interactions: multiple abrupt pushes by a person while the robot maintains balance.

example, with the robot being abruptly pushed on the upper-left shoulder. The robot was able to recover and maintain balance. With the same force control framework, the robot performed squatting, force tracking to external forces, and position tracking while keeping its balance [39].

1.2.2.3   Perception—Visual Processing and Ocular–Motor Responses

Human interaction with the external world involves the utilization of a fully integrated sensory system. To deal and interact with the external world, our system is equipped with microphones for hearing and video cameras for seeing (two eyes for stereo vision processing; two cameras per eye—foveal and peripheral camera—to mimic foveated vision of biological systems). The peripheral cameras provide a wide visual view of the environment whereas the foveal cameras provide a more detailed view of a smaller portion of the world.

Results from neuroscience research show that visual information is transferred along a number of pathways (e.g., magnocellular pathway, parvocellular-blob pathway, and parvocellular-interblob pathway) and that visual processes are executed in well-defined areas of the brain. Visual perception results from interconnections between these partly separate and functionally specialized systems. Such a processing system is much too complex to be implemented onboard. The distributed architecture described earlier was designed to allow remote execution of such processing models. It allows us to organize vision processing in a brainlike manner—serially, in parallel with feedforward connections as well as feedback connections—forming networks of interconnecting processing elements.

As a first example that tests the proposed distributed architecture we developed a visual attention system for CB based on the biologically motivated model described in References [40, 41]. This visual attention model exhibits a distributed processing architecture, which is quite typical for the processing of information in the brain. The original visual stream is subdivided into several processing streams associated with different features (color, intensity, orientation, motion, and disparity were used) that are in part independent of each other. Farther down the processing stream the

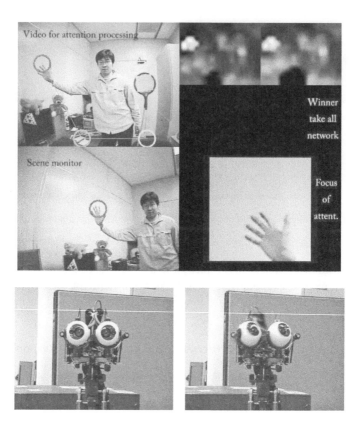

FIGURE 1.22   Visual attention. The robot actively saccades to visually acute movement. See color insert.

results must be integrated and synchronized to produce a global saliency map suitable for the selection of the focus of attention. Once the focus of attention is selected, a feedback connection is needed to suppress visual processing during the saccade. The developed architecture allows us to distribute visual attention processing among eight PC workstations and provides us with means to realize the feedforward and feedback connections. One issue that became noticeable when implementing the attention system was that the processing time and consequently the latency and frame rate can vary across the visual streams. This led us to develop and test a number of synchronization schemes that enable time synchronization of visual streams running at various frequencies and latencies [31]. The result of the visual attention system is shown in Figure 1.22.

We also experimented with ocular–motor responses similar to those of humans. The following biological ocular–motor control schemes have been incorporated into our system:

- Vergence: Minimizing target disparity by symmetric eye movement
- Saccadic eye motion: Quick knee–jerk–type eye movements to redirect gaze

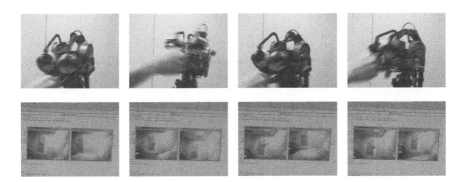

FIGURE 1.23   Vestibulo-ocular reflex (VOR). Sudden head movements while CB fixated on a target (a tennis ball); in this experiment, stability control was applied to the left eye only.

- Vestibulo-ocular reflex (VOR): Gaze stabilisation by compensating for externally induced head movements (as shown in Figure 1.23)
- Eye movements: Coupled with head movement
- Saccade: Followed by smooth pursuit of a target in an integrated control environment

These motions form the basis for the realization of active higher-level visual processes.

1.3 CONCLUSION

This chapter presented a research paradigm that brings a new direction in robotics that draws on neuroscience in the development of engineering systems, namely humanoid robotic systems.

A highly integrated humanoid robotic system, CB, developed for studies of humanlike information processing in dealing with the real world was presented. The hardware and software architectures of the overall system were presented. Three aspects of our system were discussed: (1) it provides sufficient power for walking; (2) its force controllability; and (3) it has a fully supported perceptual system, with a distributed architecture to emulate brainlike processing of sensory information.

The humanoid robot CB, we believe, is the first humanoid robotic system that is capable of full-body force-based (compliant) control. It has a similar range of motion, with a similar configuration (50 degrees of freedom), and has similar performance to that of humans. Additionally, its perceptual system attempts to mimic that of humans. This characteristic makes CB suitable as a research tool to investigate the neuroscience of information processing in the brain. This is because, we believe, a sophisticated system such as CB will impose the appropriate constraints by placing our exploration within the context of human interactions within human environments.

ACKNOWLEDGMENT

The work conducted here was partly supported by the Keihanna branch of the National Institute of Communication Telecommunication (NiCT), Japan. Additional support was gathered from the Japan Science Technology Agency (JST), under the International Cooperative Research Project (ICORP) project, Computational Brain. The contributions of the funding bodies and the research team are greatly appreciated. In addition, we wish to acknowledge the continuous support of members of SARCOS Research Corporation.

REFERENCES

1. Gregory, R. *Mind in Science*. Cambridge, UK: Cambridge University Press, 1981.
2. Kuroki, Y., Ishida, T., Tamaguchi, J. I., Fujita, M., and Doi, T.T. A small biped entertainment robot. In *Proceedings of IEEE-RAS International Conference on Humanoid Robots (Humanoids 2001)*, Tokyo, 2001.
3. Hollerbach, J.M. and Jacobsen, S.C. Anthropomorphic robots and human interactions. In *Proceedings of the First International Symposium on Humanoid Robots* (HURO'96), Tokyo, 1996.
4. Hirai, K., Hirose, M., Haikawa, Y., and Takenaka, T. The development of honda humanoid robot. In *Proceedings of the 1998 IEEE International Conference on Robotics and Automation*, Leuven, Belgium, 1998.
5. Sakagami, Y., Watanabe, R., Aoyama, C., Matsunaga, S., Higaki, N., and Fujimura, K. The intelligent ASIMO: System overview and integration. In *Proceedings of IEEE/RSJ International Conference on Intelligent Robots and Systems*, Lausanne, 2002.
6. Asfour, T., Berns, K., and Dillmann, R. The humanoid robot ARMAR. In *Proceedings of the Second International Symposium on Humanoid Robots*, 1999.
7. Asfour, T., Berns, K., and Dillmann, R. The humanoid robot ARMAR: Design and control. In *Proceedings of IEEE-RAS International Conference on Humanoid Robots (Humanoids 2000)*, Boston, 2000.
8. Bischoff, R. Advances in the development of the humanoid service robot HERMES. In *Proceedings of the Second International Conference on Field and Service Robotics*, Pittsburgh, 1999.
9. Bischoff, R. Natural communication and interaction with humanoid robots. In *Proceedings of the Second International Symposium on Humanoid Robots*, 1999.
10. Jung, H.-W., Seo, Y., Ryoo, M.S., and Yang, H.S. Affective communication system with multimodality for a humanoid robot, AMI. In *Proceedings of Humanoids*, 2004.
11. Inoue, H. et al. HRP: Humanoid robotics project. In *Proceedings of Humanoids 2000*, Boston, 2000.
12. Inoue, H. A platform-based humanoid robotics project. In *IARP First International Workshop on Humanoid and Human Friendly Robotics*, Tsukuba, Japan, 1998.
13. Hirukawa, H., Kanehiro, F., Kaneko, K., Kajita, S., Fujiwara, K., Kawai, Y., et al. Humanoid robotics platforms developed in HRP, *Robot. Auton. Syst.* **48**:165–175, 2004.
14. Ambrose, R.O., Aldridge, H.A., Askew, R.S., Burridge, R.R., Bluethmann, W., Diftler M.A., et al., Robonaut: NASA's space humanoid. *IEEE Intell. Syst.* **15**:57–63, 2000.
15. Brooks, R.A. The Cog project, *Robot. Soc. Japan*, **15**:968–970, October 1997.
16. Brooks, R.A., Breazeal (Ferrell), C., Irie, R., Kemp, C.C., Marjanovi, M., Scassellati B., et al. Alternative essences of intelligence. In *National Conference on Artificial Intelligence 1998-1999*, 1998.

17. Scassellati, B. Eye finding via face detection for a foveated, active vision system. In *Proceedings of the National Conference on Artificial Intelligence, 1998-1999*, 1998.

18. Brooks, R.A., Breazeal, C., Marjanovi, M., Scassellati, B., and Williamson, M.M. The Cog project: Building a humanoid robot. In *Proceedings of IARP First International Workshop on Humanoid and Human Friendly Robotics*, Tsukuba, Japan, 1998.

19. Kawamura, K., Wilkes, D.M., Pack, T., Bishay, M., and Barile, J. Humanoids: Future robots for home and factory. In *Proceedings of the First International Symposium on Humanoid Robots*, 1996.

20. Kuniyoshi, Y. and Nagakubo, A. Humanoid interaction approach: Exploring meaningful order in complex interactions. In *Proceedings of the International Conference on Complex Systems*, New Hampshire, 1997.

21. Nagakubo, A., Kuniyoshi, Y., and Cheng, G. Development of a high-performance upper-body humanoid system. *Adv. Robot.* **17**:149–164, 2003.

22. Cheng, G., Nagakubo, A., and Kuniyoshi, Y. Continuous humanoid interaction: An integrated perspective – Gaining adaptivity, redundancy, flexibility – In one. *Robot. Auton. Syst.* **37**:163–185, November 2001.

23. Sandini, G. Babies and baby-humanoids to study cognition. In *Proceedings of the BICA*, 2012.

24. Tistarelli, M. and Sandini, G. On the advantages of polar and log-polar mapping for direct estimation of time-to-impact from optical flow. *IEEE Trans. Patt. Anal. Mach. Intell.* **15**:401–410, 1993.

25. Sandini, G., Metta, G., and Vernon, D. RobotCub: An open framework for research in embodied cognition. In *Proceedings of the IEEE International Conference on Humanoid Robots*, Santa Monica, 2004.

26. Kozima, H. Infanoid: An experimental tool for developmental psycho-robotics. In *Proceedings of the International Workshop on Developmental Study*, 2000.

27. Atkeson, C.G., Hale, J.G., Pollick, F., Riley, M., Kotosaka, S., Schaal, S., et al. Using humanoid robots to study human behavior, *IEEE Intell. Syst. Appl.* **15**:46–56, July/August 2000.

28. Ishiwata, Y. How to use Linux as a real-time OS. *Interface,* pp. 109–118, November 1999.

29. Hale, J.G., Hohl, B., Hyon, S.-H., Matsubara, T., Martin, E., and Cheng, G. Highly precise dynamic simulation environment for humanoid robots with a transparent interface. *Adv. Robot.* **22**:1075–1105, 2008.

30. Essen, D.C.V. and DeYoe, E.A. Concurrent processing in the primate visual cortex. In M.S. Gazzaniga *(Ed.), The Cognitive Neurosciences.* Cambridge, MA: MIT Press, 1994, pp. 383–400.

31. Ude, A.v., Wyart, V., Lin, M., and Cheng, G. Distributed visual attention on a humanoid robot. In *Proceedings of IEEE International Conference on Humanoid Robots*, Santa Monica, 2005.

32. Gumpp, T., Azad, P., Welke, K., Oztop, E., Dillmann, R.U., and Cheng, G. Unconstrained real-time markerless hand tracking for humanoid interaction. In *Proceedings of IEEE-RAS International Conference on Humanoid Robots (Humanoids 2006)*, Genova, Italy, 2006.

33. Endo, G., Morimoto, J., Matsubara, T., Nakanishi, J., and Cheng, G. Learning CPG sensory feedback with policy gradient for biped locomotion for a full body humanoid. In *Proceedings of the Twentieth National Conference on Artificial Intelligence (AAAI-05)*, Pittsburgh, 2005.

34. Matsubara, T., Morimoto, J., Nakanishi, J., Sato, M., and Doya, K. Learning sensory feedback to CPG with policy gradient for biped locomotion. In *Proceedings of IEEE International Conference on Robotics and Automation*, Barcelona, 2005.

35. Nakanishi, J., Morimoto, J., Endo, G., Cheng, G., Schaal, S., and Kawato, M. Learning from demonstration and adaptation of biped locomotion. *Robot. Auton. Syst.,* **47**(2):79–91, 2004.

36. Morimoto, J., Endo, G., Nakanishi, J., Hyon, S.-H., Cheng, G., Bentivegna, D., et al. Modulation of simple sinusoidal patterns by a coupled oscillator model for biped walking. In *Proceedings of IEEE International Conference on Robotics and Automation,* Orlando, FL, 2006.

37. Winter, D.A. *Biomechaniics and Motor Control of Human Movement,* 2nd ed. New York: John Wiley & Sons, Inc., 1990.

38. Arimoto, S. *Control Theory of Non-Linear Mechanical Systems: A Passivity-Based and Circuit-Theoretic Approach.* New York: Oxford University Press, 1996.

39. Hyon, S.-H. and Cheng, G. Gravity compensation and whole-body force interaction for humanoid robots. In *Proceedings of IEEE-RAS International Conference on Humanoid Robots (Humanoids 2006),* Geno, Italy, 2006.

40. Itti, L. and Koch, C. Computational modelling of visual attention. *Nature Rev. Neurosci.* **2**:1–11, March 2001.

41. Itti, L., Koch, C., and Niebur, E. A model of saliency-based visual attention for rapid scene analysis. *IEEE Trans. Patt. Anal. Mach. Intell.* **20**:1254–1259, 1998.

2 Humanoid Brain Science

*Erhan Oztop, Emre Ugur, Yu Shimizu, and
Hiroshi Imamizu*

CONTENTS

This chapter underlines the reciprocal interaction between brain sciences and humanoid robotics. Although robots serve best for testing hypotheses about brain function, neuroscience provides engineering clues to designing better robots. Here, we present research efforts representative of both sides of this interaction, and propose the term "humanoid brain science" that embraces these works. Humanoid brain science covers a wide multidisciplinary research spectrum that exploits the intrinsic connection between robotics and biological systems that stems from the observation that both systems are required to solve similar computational problems in the very same physical environment.

2.1 BASIC ROBOTIC LESSONS FROM BEHAVIORAL EXPERIMENTATION

2.1.1 SENSORIMOTOR LEARNING

We know firsthand that we (or better, our bodies) can learn to deal with novel dynamics, let it be the control of a novel tool or the control of our body in different environments.

This is possible due to the central nervous system (CNS), which is equipped with a set of adaptive control mechanisms that keeps learning. One of the influential paradigms in understanding human motor control and its adaptation is the so-called force field paradigm [50]. In this paradigm, the human is asked to perform reaching movements while holding on to a handle. The handle is in fact a robot that can induce desired forces depending on the state of the subject's hand. With this setup it has been shown that humans learn to adapt their movements apparently to counteract perturbations in a state-dependent way. Furthermore, it is shown that this is not achieved by stiffening of the muscles, but rather by active force generation to react to the learned force field. This indicates that the CNS optimizes energy by building internal predictive models to generate forces in an anticipation of the perturbation. The debate as to what is the exact mechanism of movement generation employed by the CNS is not settled. On one side, researchers propose that the observed effects can be explained by the optimal feedback control employed by the CNS. The other, more classical, view is that the desired trajectory governing reaching movements is tracked by learning internal models (see the next subsection) to cancel out the perturbations. Yet, another, rather reconciliatory view is, that which the CNS actually does, to reoptimize the desired trajectory in the face of large perturbations [26]. From a computational view, adoption of pure optimal force control (OFC) seems computationally infeasible for robotic systems, leaving the reoptimization view the most reasonable candidate for robotic implementations.

Note also that this brief discussion only reflects one dimension of how the CNS controls movement. Other issues such as how learning ensues and is guided by the CNS are also critical. For example, it has been shown that learning behavior is influenced by motor memory, which often causes suboptimal task solutions to be adopted even with knowledge of the optimal solution [21]. Another important issue not discussed here is stiffness. Although we mentioned predictive perturbation cancellation, the CNS also actively controls stiffness, and it does this in a direction-selective way; that is, stiffening occurs only in the direction of the perturbation [19, 27]. Furthermore, for different limbs or contexts the CNS may adopt different strategies, for example, when the consequence of a failure is fatal, such as in posture control [4]. Often in most robotics work unstable environments are avoided; when interaction with the environment is considered, robust control approach is usually used to obtain interaction stability with a fixed control structure [13]. Humans use a different strategy to deal with instability [10] and control the impedance of the end-effector to keep a stability margin within which error minimization and energy minimization can be performed. These ideas now have begun to be expressed in computational learning models [18] that are applied to actual robotic platforms with variable impedance actuators [13].

Another note is in order here. The results presented here are mainly derived from reaching experiments; although the data provide many insights, care must be taken when extrapolating the data to general motor control. Our preliminary experiments involving full body posture control experiments indicate that humans may use a different strategy for counteracting postural perturbations compared to reaching perturbations [3].

2.2 BASIC ROBOTICS LESSONS FROM NEUROSCIENCE

2.2.1 MIRROR NEURONS AND BODY SCHEMA

A set of neurons in the ventral premotor area of macaque monkeys has been found to respond to the observation of goal-directed movements performed by another monkey or an experimenter (e.g., precision or power grasping) for actions more or less similar to those associated with the motor activity of the neuron [20, 45]. These neurons are called mirror neurons (MNs) due to their dual response property. Some of these neurons were further found to respond to the characteristic sound of specific actions (e.g., breaking a peanut) [29] or to actions of which the last part of execution is hidden from view (e.g., [53]. The initial excitement in robotics was that mirror neurons justified the imitation learning scenarios, which involve a shortcut to access the demonstrator's motor parameters. Yet, this did not solve the real problem for robotics. The observed visual input still needed to be converted into motor parameters through a computational process in a realistic deployment. Therefore this was not the key insight that robotics obtained from neuroscience. It was rather the idea of reusing motor control mechanisms for perception. This created a big paradigm shift in robotics and created new research areas that see perception and action tightly coupled. The exact mechanism as to how the brain combines those two is far from clear, and consequently the robotic implementations are rather limited, but they do appear to be in the right direction [16, 52].

As a final note, in spite of the views on their involvement in imitation and other higher-level cognitive functions, it is not at all clear what the exact role of *mirror neurons* in monkeys and *mirror areas* in humans is [41]. The location of F5 mirror neurons suggests that they were initially used for motor control, which may later have taken over cognitive tasks in the course of evolution. In this spirit, it has been proposed that mirror neurons could be involved in forward modeling [44] and inverse modeling [16]. Two simple mechanisms that may lead to mirrorlike responses are self-observation and autoassociation where the motor representation of the action and the representation of the visual effect generated by it are associated through a Hebbian-like mechanism [28]. This mechanism may also bootstrap simple imitation or mimicry as discussed in the next section.

2.2.2 BODY SCHEMA

As with robotic systems, biological systems must know about the state and intrinsic properties of their bodies in order to move, avoid collision, and manipulate objects in their environment. The neural representation of the body in the brain for this purpose is referred to as the body schema. It is an integrated representation that involves visual, somatosensory, and proprioceptive modalities. It appears to operate in body-part–centered reference frames and exhibits significant plasticity. The key property of the body schema is that it is constantly updated with the available sensory input [23], and it is quite easy to create striking illusions with simple manipulations. An example of such an illusion is the rubber hand illusion in which a subject views an artificial rubber hand being stroked while simultaneously being applied a similar stroking motion to his own hand positioned out of view. The visual and tactile contingency then creates a feel of touch when the subject observes someone touch the rubber hand, even when his own hand is untouched [8]. Neurophysiological experiments with behaving monkeys have shown that primates are endowed with a plastic representation of body parts, specifically hands, which are expanded when a tool (a rake) is used to manipulate objects in the environment of the monkey. To be more concrete, specific parietal area neurons that encode the space on and around the hands of the monkey change their response properties when a tool is grasped by the monkey, and start to encode it as if it were an extension of the monkey's arm [25, 38, 32]. Taken together this underlines the adaptive nature of the *body schema*, the representation of the body in the human brain, and how this adaptation can work at very short timescales. Terms that are closely related to body schema are *body image* and *agency* [17]. The former indicates a conscious knowledge of the body, whereas the latter refers to the sensation of being the owner of the observed action (i.e., being the agent who caused the action). As the neuroscientific bases of these faculties are being uncovered the straightforward path for robotics is to utilize current knowledge for designing robot architectures that parallel the working principles of the brain. A rather less adopted path is to exploit the human capacity to represent tools as body parts for robot skill generation, about which we have more to say in the following sections of this chapter.

2.3   TESTING HYPOTHESES WITH ROBOTS: FROM SELF-OBSERVATION TO IMITATION

Imitation receives considerable attention from both human sciences and robotics. For psychology and neuroscience it holds the key for other high-level cognitive abilities such as language and the theory of mind [33]. For robotics, it offers a unique opportunity to program robots without requiring experts. Imitation covers a set of behaviors sharing the common factor of transforming an observed action into action, which may widely vary in terms of the type and feature of the action imitated [12]. Imitation may mean to understand the goal of the demonstrator and perform an action, which may or may not be similar to the observed one, to achieve the goal. On the other hand, it may mean a faithful reproduction of the observed action. In general the question is what and how to imitate [37, 49, 6]. Therefore, imitation covers a continuum of behaviors ranging from simple, automatic, and involuntary action contagion to intentional imitation, and emulation [11].

We contend that the human brain is endowed with several levels of imitation strategies. We maintain that at the lowest level is the action mimicry, the simple imitation of an observed act without assessing the goal or the intention behind it. A recurring theme in our past modeling studies was self-observation. Self-observation allows for the development of internal models that can be utilized for inferring other people's goals [44]. It may also underlie the initial formation of mirror neurons [39]. In terms of mimicry we asked whether it could be bootstrapped in an anthropomorphic robotic agent, who can control and observe its own hand similar to a developing infant.

It is often assumed that the imitation is innate, referring to the influential paper of Meltzoff and Moore [34] reporting neonatal imitation, although some criticism exists with respect to the results and the interpretation of the findings [1]. Initial imitative abilities may have evolved in order to induce a social bond between a mother and her baby; however, neonatal imitation is far from explaining the full range of imitative skills observed in primates. We hold that basic forms of imitation emerge as a result of interaction with the environment through self-observation, and simple neural circuits sustain this imitation. A classical view related to synaptic plasticity and learning in the brain is Hebbian learning ("When an axon of cell A is near enough to excite a cell B and repeatedly or persistently takes part in firing it, some growth process or metabolic change takes place in one or both cells such that A's efficiency, as one of the cells firing B, is increased." Hebb, D.O. (1949). The Organization of Behavior.). From a computational point of view this learning mechanism is not very useful in its original form. However, slight relaxations to the original statement allow for construction of the so-called (auto-) *associative memories* [22]. The key point of an autoassociative memory is that a partial representation of a stored pattern can be used to retrieve the whole. Assuming that similar mechanisms are at work in the cerebral cortex, we can envision a neural mechanism that can generate simple imitative behavior, that is, mimicry. When the agent generates motor commands for movement, the representation of this command and the sensory representation of the effects of the movement can be associated through Hebbian-like learning. At a later time when the system encounters a stimulus that partially matches the stored items, for example, if only the vision input is available, the associated motor command can be retrieved and hence used to mimic the observed movement thanks to autoassociation. This line of thought has also been explored by others employing different neural architectures [30].

We tested this hypothesis by adopting an extension of the Hopfield network model that captures the Hebbian learning paradigm, and creating a robotic setup that involves a 16-degrees-of-freedom robot hand and a video camera [40, 12]. The camera modeled the eyes of the agent and was used to implement self-observation. The robot was given a discrete set of movements, which involved the extension and flexion of four fingers (Figure 2.1).

Although in a biological setting, learning and testing take place in an interleaved manner, for simplicity it was assumed that the imitation system was in either the learning or the testing phase. The system could learn by self-observation or from pure human demonstration. As we are interested here in self-observation, we briefly present the case for self-observation learning. In the learning phase the video camera was directed at the robotic hand while the robot executed finger movements and "watched" itself. The motor patterns and the resulting processed video images were then associated using an associative neural network. This network stored, in a distributed manner, the retinal patterns of the hand and the motor commands that yielded the respective patterns.

The network was composed of 640 artificial neurons that received inputs from the retinal pixels and the units representing the motor commands. Each artificial neuron was connected with all other neurons through multiplicative synapses. After learning, the system could be tested for imitative abilities. In this mode, the outputs of the network nodes corresponding to the visual inputs are initialized from the retinal input that may include a human hand. The outputs of the motor-related nodes are set arbitrarily as either 0 (no activity) or 1 (full activity). Then the network evolves

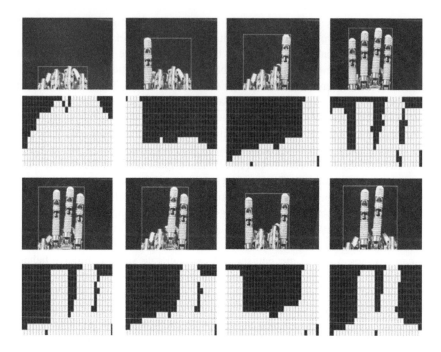

FIGURE 2.1 The robot hand postures used are shown as seen by the camera and the modeled retina. (Courtesy of Chaminade, Oztop et al. 2008. With permission.)

according to the modified Hopfield rule using the weights computed during learning and settles to an attractor. The motor part of the converged attractor then is used to drive the robot fingers. Because the training was purely based on the self-observation of the robot, the imitation of an observed human hand posture, which is quite different from the robot hand, was the key evaluation measure with respect to the system's ability to bootstrap mimicry. The tests showed that most of the human hand postures were reliably imitated (the single exception being full hand extension, probably due to the large span of the robot hand). Figure 2.2 shows the posture imitation capability derived through self-observation.

In sum, this work emphasizes, via exemplifying, that robotic systems are well suited for testing hypotheses about human visuomotor development when combined with biologically realistic neural networks and sensorimotor systems comparable to their biological counterparts. In particular, the work lends support, both computationally and neuroscientifically, to the hypothesis that simple imitation of hand actions is bootstrapped through self-observation.

2.4   FROM BRAIN TO ROBOT SKILL GENERATION

When a new skill is desired for a given robot, an expert (or an expert team) is needed to implement this skill on the robot. If robots were able to imitate what humans demonstrate, obtaining novel robot behaviors would become simple. Considerable

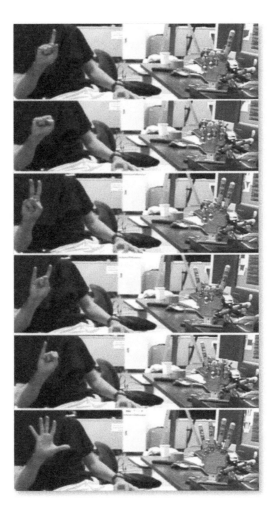

FIGURE 2.2 Gifu hand imitating human hand postures. The imitation system was boot-strapped using self-observation.

effort in robotics is spent on developing systems that can learn by observing demonstrated actions [48, 9, 15, 24, 49, 5, 7, 6]. If the skill is demonstrated visually, the first task to be tackled is a nontrivial computer vision problem (see Azad [2] for a recent review). When the state of the demonstrator can be measured, the problem can be reduced to the mapping of the observed measurements to the target robot. However, this is not always trivial due to kinematic and dynamic differences between the observer and the demonstrator. Moreover, the mapping must be designed by a robotics expert. In direct teaching [31, 46, 47], that is, by actively moving the robot joints to obtain the desired behavior, some of these problems are circumvented. However, this approach is not applicable for complex tasks that may include nonnegligible dynamics.

To overcome these difficulties, we proposed a new paradigm, dubbed "robot skill generation using human sensorimotor learning" (RSHL), a paradigm that overcomes these difficulties [42, 43]. This paradigm taps into the human motor control system by establishing an intuitive teleoperation system, where a human operator learns to control the joints of a robot to perform a given task. The impetus for the development of these paradigms is the notion that by practice the robot will be incorporated into the body schema of the operator, as in the rake-using monkeys [25]. A robot could simply be considered as a tool to perform a given task, which the user would learn to control by practice. This paradigm places the initial burden of learning on the human operator, but ultimately allows the robot to acquire the target skill without human guidance. This may sound similar to the direct teaching mentioned earlier, with one critical difference. In direct teaching, the cognitive system of the human is guiding the robot, whereas in the sensorimotor learning paradigm the human motor system is actively engaged in controlling the robot. Involvement of the motor system makes a huge difference as it incorporates dedicated circuits for the control and learning of the dynamic characteristics of the limbs and external objects (Imamizu 2010; Shidara, Kawano et al. 1993). Unlike the sensorimotor system, the cognitive system is not oriented toward minute-to-minute error correction and real-time feedback processing critical for control tasks.

In the RSHL paradigm the construction of a motor skill takes place in two phases: (1) a human operator learns to perform the task using the robot as a tool, and (2) after phase 1, the signals received by the the robot during the human task performance are associated with machine-learning algorithms together with the robot and environment state, so as to yield an independent controller. Here the emphasis is on the data rather than the specific machine learning method-adopted. The amount of data, the ease of their acquisition and the rate at which they can be produced is the benefit brought by the RSHL paradigm. With this approach following the initial ball-swapping skill [42, 43, 51], stable reaching for a small humanoid robot [4], and basic grasping with an anthropomorphic robot were achieved [35].

2.4.1 Ball Swapping

For the ball-swapping task, a 16-degree-of-freedom (DOF) robot hand (Gifu Hand III, Dainichi Co. Ltd., Japan) placed nearly parallel to the ground in a palm-up orientation was used. The task was to move the fingers (only 12 DOFs were used) of the robot hand so that the balls located on the fingers swap their locations without rolling down. The human operator finger motion was captured and mapped to the robot in real-time. The operator had to learn how to control the robot fingers, so that that the balls were swapped without dropping. The operator "invented" the strategy to jerk the ball with small fast finger movements rather than sliding it over the surface of the fingers (see Figure 2.3).

This strategy, even though different from when the task is executed in natural conditions, allowed robust ball swaps with the robot under human control. The performance of the operator is then used to generate autonomous open loop [42, 43] and feedback [51] ball-swapping skills. For the former the motor commands were not associated with the state, so there was no machine learning; but a simple

FIGURE 2.3 Frames representing the ball-swapping skill obtained using the human senso-rimotor learning paradigm. The numbers in the top left corner of the images indicate the chronological order of the frames. (Adopted from Moore and Oztop 2012. With permission.) See color insert.

playback was involved. For the latter, unsupervised kernel regression was used to map the robot and ball state to motor commands. Figure 2.3 shows a set of frames from the obtained autonomous ball swapping execution.

One of the current efforts is the application of the RHSL paradigm to walking and posture control of a full-sized humanoid robot. Directed toward this aim, first, a simpler, full-body control task was targeted, more precisely, the statically stable reaching using a small humanoid robot. Technically, the problem is an inverse kinematics problem with a stability constraint (i.e., keeping the ground-projected center-of-mass (COM) point in the feet support polygon). For this task we modified one dimension of the paradigm, namely how the robot state was fed back to the operator. Instead of asking the operator to observe the robot and control it while judging its stability, we presented a rendering of the support polygon and the ground-projected center-of-mass point on a computer display. The task of the operator was to move his hand up and down as guided by the experimenter while keeping the COM point in the support polygon. This feedback was very effective and the operator could control the robot as instructor after a short training. In the data collection period the operator generated joint angles and end-effector position information that satisfied the static stability constraint. The end-effector to joint angle mapping was then estimated by a function approximator, which yielded an inverse kinematic mapping that respected the stability constraint. Using this mapping any desired trajectory in the sagittal plane can be tracked as illustrated in Figure 2.4.

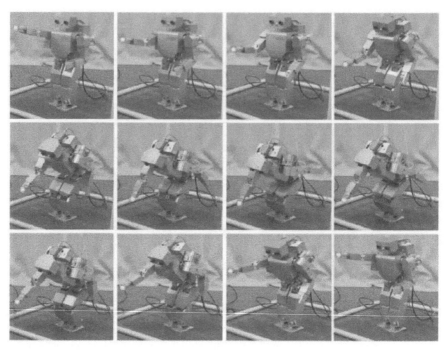

FIGURE 2.4 Frames from a statistically stable trajectory tracking obtained through the RSHL paradigm. (Adopted from Babič, Hale et al. 2011. With permission.)

2.5 FROM ROBOT SKILL GENERATION BACK TO BRAIN

The initial motivation for developing the RSHL paradigm was based on neuroscience: an anthropomorphic robot that is controlled by a human for long periods would eventually be incorporated into the body schema of the operator, enabling her to control the robot smoothly. We mentioned above that when a tool (a rake allowing the manipulation of objects out of reach) is grabbed by a trained monkey, parietal neurons with receptive fields on and around the hand expand their receptive fields to represent the tool [25, 38]. Because the human gains the ability to manipulate objects using a robot, can we project from the finding mentioned above that the body schema of a human operator expands to include the robot in the course of controlling it? And if so, what are the properties of this body schema expansion? Does the control of all kinds of robot morphology induce similar changes in the brain? To investigate these questions we designed an fMRI experiment. Our hypothesis was that the anthropomorphicity of the robot to be controlled would be critical to how it was represented in the brain. We assumed that, if the robot were anthropomorphic and the control required from the subject compatible with the robot, the body schema would easily subsume the robot. Otherwise, other brain mechanisms such as internal models would be employed more heavily.

2.5.1 ROBOTS

Two different (simulated) robots were used to test our hypotheses, where the subject controlled the robots using his fingers. For the anthropomorphic robot control the Gifu Hand III (Dainichi, Japan), which has five fingers and a total of 16 DOFs, was used. For the nonanthropomorphic robot control a 7-DOF robot arm (Motoman SDA10, Yaskawa, Japan) was used. In both cases the control was based on human finger movements. Both of these robots were simulated in a physics-based simulation environment and rendered on a screen that was projected onto the display unit installed in the MRI. This setup allowed subjects to see the simulated robots as they controlled them. The same display was also used to instruct the subjects on the task to be performed.

2.5.1.1 Human Control

The human subjects used the index and middle finger to control the robots. Finger joint angles were measured using an MRI-compatible data glove and mapped to either the robot hand finger angles (anthropomorphic robot) or the robot arm joint angles (nonanthropomorphic robot). Only the angles of the proximal finger joints were used. In the hand robot case, this mapping was anthropomorphically correct; that is, the index and middle fingers of the robot hand were controlled by the corresponding human fingers by sending the flexion angle obtained from the data glove to the proximal joints of the corresponding robot fingers. In the arm robot case, the human index finger controlled the ulnar rotation of the robot forearm while the human middle finger controlled the robot wrist flexion/extension. Furthermore, to help the human subject in perceiving the position and orientation of the robot arm, a 3D coordinate frame was superimposed on the end-effector of the simulated arm. The motion range of the joints for both robots was limited to $10°-80°$.

2.5.1.2 Task

The task was to move the robots to an indicated configuration. In the robot hand case the target configuration could be achieved by bringing robot fingertips to their desired positions that were drawn as two small spheres in different colors (see Figure 2.5 middle panel).

In the robot arm case, the desired configuration was achieved by bringing the robot arm end-effector to a desired position (or orientation, as both necessarily require the same output from the subject). The target was rendered as a 3D coordinate frame whose center and orientation corresponded with the desired position and orientation of the arm, respectively (see Figure 2.5, right panel). The desired targets for both robots required exactly the same finger movements from the subject.

2.5.1.3 Experimental Design

A conventional block design was used for the fMRI experiment. One experimental session was composed of 32 blocks, each containing 16 robot control (execution) or observation tasks, in the following referred to as trials. Four types of blocks

FIGURE 2.5 On the left, the human hand with a data glove controls the simulated robot hand. In the middle, a snapshot from the anthropomorphic control experiments is shown. The subject uses his index and middle fingers to control the corresponding robot fingers, so that the robot fingertips reach the targets presented as colored spheres. On the right, the subject's goal is to move the end-effector of the robot arm to the presented target position using his fingers through nonanthropomorphic control. 3D coordinate frames are attached to the robot arm and target position to aid the subject's 3D position perception. In addition, the dashed circles show the rotating joints of the robot arm.

were sequentially repeated within one session: (1) anthropomorphic control execution (AN-EXE), (2) anthropomorphic control observation (AN-OBS), (3) nonanthropomorphic control execution (NAN-EXE), and (4) nonanthropomorphic control observation (NAN-OBS).

In the execution blocks the task of the subject was to bring the robots into the desired configurations by moving their index and middle fingers. The configurations were the result of two angle choices for both robots, which were selected pseudorandomly (without repetition) in the range of $[30°–60°]$.

During an observation block subjects were asked to relax and watch the presentation on the screen, where the previous execution block was played back; targets as well as the robot movements were presented exactly as conducted in the previous block. These observation blocks were introduced to remove the vision-related brain activity in the analysis phase.

Prior to scanning, the subjects were trained until the average angular error between the target and the actual robot joint angles at the end of each trial fell below a fixed threshold ($<5°$). After the subjects became sufficiently accustomed to the task, the subjects were taken into the MRI scanner and asked to go through the same experimental sessions with the addition of the observation conditions mentioned above. The fMRI experiment paradigm adopted is summarized in Figure 2.6. The subjects repeated the blocks AN-EXE, AN-OBS, NAN-EXE, NAN-OBS eight times. In each block, 16 target positions were presented to the subject for 3 seconds, during which the subjects were required to bring the robots into the desired configurations. We collected data from four subjects but here report on only one subject as experimentation at the group level is still in progress.

2.5.1.4 Data Collection and Analysis

fMRI data were obtained for the whole brain using an echo planer image (EPI) sequence (TR = 3 s, echo time = 0.03 s, dimensions: $50 \times 64 \times 64$, voxel size: $3 \times 3 \times 3$ mm, no gap) using a 3 Tesla scanner (Magnetom 3T Trio, Siemens, Germany).

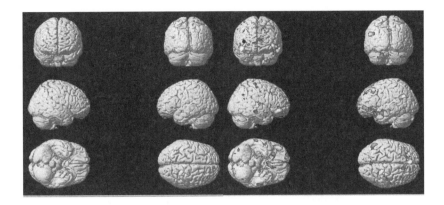

FIGURE 2.6 AN-NAN (left) and NAN-AN (right) contrasts for a single subject (p<0.005).

During each session 619 functional volumes were obtained. T1-weighted structural images (matrix size: 192 × 256 × 256, voxel size 1 × 1 × 1 mm, flip angle = 80°) were obtained with a gradient echo sequence for further usage in fMRI analysis. T2-weighted structural images were obtained for later co-registration of the functional data.

The fMRI data were processed with SPM8 (Wellcome Department of Cognitive Neurology, London, UK; http://www.fil.ion.ucl.ac.uk/spm). We discarded the first three volumes of functional images in each session to allow for T1 equilibration, and then aligned the data to the first remaining volume. The T2 structural volume was co-registered to the mean EPI. Both the T2 and EPIs were then registered to the structural T1 volume. T1 volume and co-registered EPIs were then normalized using the Montreal Neurological Institute (MNI; Montreal, Canada) reference brain. Data were smoothed spatially with a Gaussian kernel of 6 × 6 × 6 mm full width at half maximum (FWHM). A high-pass filter with a cut-off frequency of 128 s was applied.

Statistical analysis was performed for each subject. Boxcar functions modeled execution blocks and observation blocks. They were convolved with the canonical hemodynamic response function in SPM8 to yield regressors in a general linear model. A parameter was estimated for each regressor using the least-squares method. T-statistics were used for comparison of the estimated parameters of AN = (AN-EXE – AN-OBS) and NAN = (NAN-EXE – NAN-OBS). Two contrasts of AN-NAN and NAN-AN then yielded a t-value for each voxel. A threshold of P < 0.005 was used in obtaining the activation maps shown in this chapter. The AN-NAN contrasts were performed to locate regions that may be more highly activated during anthropomorphic control than during nonanthropomorphic control. Similarly, the NAN-AN contrast was evaluated to locate regions that may contribute more in nonanthropomorphic control.

FIGURE 2.7 The experimental paradigm. Anthropomorphic and nonanthropomorphic control correspond to hand and arm control, respectively. In the visual blocks, the subjects do not move, but watch a recording of the previous execution block including the robot movements.

2.5.1.5 Results

Even though task performance was comparable in both AN and NAN conditions, it appears that the NAN condition required more mental effort as indicated by the higher activation in the NAN case compared to AN case (Figure 2.6). Furthermore, high prefrontal activation in the NAN-AN contrast indicates task difficulty in the NAN condition. A more interesting result is that the NAN condition engaged the angular gyrus, which is involved in the detection of a mismatch between the intended and actual movement leading to a loss of "action ownership" (agency). This is interesting because in terms of behavioral performance these two robot tasks did not differ significantly. This may suggest that the anthropomorphic mismatch in the NAN conditions can create a loss of agency in controlling an arm with the fingers when compared with the control of fingers with fingers. It is likely that with anthropomorphicity, the threshold for agency rejection may be elevated.

In normal reaching and pointing movements, superior parietal regions are involved in programming the movement according to extrinsic spatial information [14]. The activation in this region suggests that the subject controlled the robot fingers as if they were the subject's own fingers, thereby supporting the hypothesis that the hand robot was incorporated into the body schema. The occipital activity in the AN-NAN contrasts may reflect the necessary fine control around the target points, which relies on detailed visual information. In the NAN case the global position of the robot arm may be more important than precise local information.

Another observation in the NAN-AN contrast is the heavy activity in the cerebellum. This may indicate that internal models representing the robot arm are in heavy use whereas in the robot hand case much of the load of the internal models might be taken over by the parietal body schema representation.

It must be noted again that the data here are from a single subject; the results should therefore only be taken as indicatory. The findings must be shown to generalize to more subjects through additional experiments, and the task difficulty must be balanced better so that the prefrontal load in both the AN and NAN cases are comparable.

2.6 CONCLUSION

We propose humanoid brain science as an umbrella covering not only engineering and robotics studies that make use of knowledge from neuroscience and human behavior, but also neuroscientific and behavioral research that is based on or derived from robotics and computational fields. As examples of humanoid brain science research, we first laid out the lessons learned from human reaching studies and mirror neuron findings. We then explained how robotics can serve as a testbed for theories about behavior and brain function through a model of mimicry. Finally we introduced the HLRS paradigm that underlines how robot technology can benefit from the study of human sensorimotor learning, and presented our initial efforts in investigating the brain mechanisms using an HLRS paradigm.

The human behavioral findings indicate that multiple adaptive control mechanisms might be the key for obtaining humanlike dexterity and robustness in robots. Complementary to this, somewhat against the classical robotics view of first-perceive-then-act strategy, neuroscience showed that perception (sensing) and action (control) should be treated in a single framework. These two features of sensorimotor control and learning were fundamental in developing the HLRS paradigm, which allows for generation of robot skills by laymen.

The robot skills obtained by HLRS indicate that exploiting the human capacity to subsume external agents into the body schema is an efficient robot skill generation scheme that does not require expert knowledge. When thought of in terms of robot programming, cutting down the human sensorimotor learning time is critical, especially if we entertain the dream of having robots accompany us in our daily lives. Once the RSHL paradigm is well established both in terms of technology and its relation to the human body and brain, the way manufacturers program their robots will change drastically. The emergence of a new employment area can be speculated upon that calls for robot trainers who would not need to be engineers at all.

The current work of functional brain imaging of RSHL and its extensions will have two types of impacts. In terms of robotics, the results will be factored into robot design allowing effective and robust programming of robots and prosthetics. In terms of neuroscience, the findings will shed light on the neural mechanisms of body schema and body image, which may help ameliorate impairments related to those neural faculties. In particular, the predictions and questions posed by the brain imaging of RSHL will stimulate neuroscientists to conduct further experiments, establishing a virtuous cycle of investigation that contributes, at the same time, to robotics technology and neuroscience of body schema, body image, and sensorimotor learning in general.

ACKNOWLEDGMENT

This research was partially supported by a contract in H23 with the Ministry of Internal Affairs and Communications, Japan, entitled "Novel and innovative R&D making use of brain structures."

REFERENCES

1. Anisfeld, M. Only tongue protrusion modeling is matched by neonates. *Develop. Rev.* **16**(2):149–161, 1996.
2. Azad, P. *Visual Perception for Manipulation and Imitation in Humanoid Robots.* Heidelberg: Springer-Verlag, 2009.

3. Babic, J., Oztop, E., et al. (2011). *Full Body Motor Adaptations to Dynamic Postural Perturbations*. PMCVIII, Progress in Motor Control VIII, Cincinnati, OH.

4. Babiè, J., Hale, J.G., et al. (2011). Human sensorimotor learning for humanoid robot skill synthesis. *Adapt. Behav.* **19**(4):250–263.

5. Bentivegna, D.C., Atkeson, C.G., et al. (2004). Learning tasks from observation and practice. *Robot. Auton. Syst.* **47**(2–3):163–169.

6. Billard, A. and Siegwart, R. (2004). Robot learning from demonstration. *Robot. Auton. Syst.* **47**(2–3):65–67.

7. Billard, A., Epars, Y., et al. (2004). Discovering optimal imitation strategies. *Robot. Auton. Syst.* **47**(2–3):69–77.

8. Botvinick, M. and Cohen, J. (1998). Rubber hands feel touch that eyes see. *Nature* **391**(6669):756.

9. Breazeal, C. and Scassellati, B. (2002). Robots that imitate humans. *Trends Cogn. Sci.* **6**(11):481–487.

10. Burdet, E., Osu, R., et al. (2001). The central nervous system stabilizes unstable dynamics by learning optimal impedance. *Nature* **414**(6862):446–9.

11. Byrne, R.W. and Russon, A.E. Learning by imitation: A hierarchical approach. *Behav. Brain Sci.* **21**(5):667–84; discussion 684–721, 1998.

12. Chaminade, T., Oztop, E., et al. (2008). From self-observation to imitation: Visuomotor association on a robotic hand. *Brain Res. Bull.* **75**(6):775–84.

13. Chenguang, Y., Ganesh, G., et al. (2011). Human-like adaptation of force and impedance in stable and unstable interactions. *Robot. IEEE Trans.* **27**(5):918–930.

14. Culham, J.C., Cavina-Pratesi, C., et al. (2006). The role of parietal cortex in visuomotor control: What have we learned from neuroimaging? *Neuropsychologia* **44**(13):2668–84.

15. Demiris, Y. and Hayes, G. (2002). Imitation as a dual-route process featuring predictive and learning components: A biologically-plausible computational model. *Imitation in Animals and Artifacts*. K. Dautenhahn and C. Nehaniv (Eds). Cambridge, MA: MIT Press.

16. Demiris, Y. and Johnson, M. (2003). Distributed, predictive perception of actions: A biologically inspired robotics architecture for imitation and learning. *Connect. Sci.* **15**(4).

17. Farrer, C. and Frith, C.D. (2002). Experiencing oneself vs another person as being the cause of an action: The neural correlates of the experience of agency. *Neuroimage* **15**(3):596–603.

18. Franklin, D.W., Burdet, E., et al. (2008). CNS learns stable, accurate, and efficient movements using a simple algorithm. *J. Neurosci.* **28**(44):11165–73.

19. Franklin, D.W., Liaw, G., et al. (2007). Endpoint stiffness of the arm is directionally tuned to instability in the environment. *J. Neurosci.* **27**(29):7705–16.

20. Gallese, V., Fadiga, L., et al. (1996). Action recognition in the premotor cortex. *Brain* **119**:593–609.

21. Ganesh, G., Haruno, M., et al. (2010). Motor memory and local minimization of error and effort, not global optimization, determine motor behavior. *J. Neurophysiol.* **104**(1):382–390.

22. Hassoun, M. *Associative Neural Memories: Theory and Implementation*, Cambridge, UK: Oxford University Press, 1993.

23. Holmes, N. P. and Spence, C. (2004). The body schema and the multisensory representation(s) of peripersonal space. *Cogn. Process.* **5**(2):94–105.

24. Ijspeert, A., Nakanishi, J., et al. (2003). Learning attractor landscapes for learning motor primitives. In S. Becker, S. Thrun, and K. Obermayer (Eds.) *Advances in Neural Information Processing Systems*. Cambridge, MA: MIT Press. **15**:1547–1554.

25. Iriki, A., Tanaka, M., et al. (1996). Coding of modified body schema during tool use by macaque postcentral neurones. *Neuroreport* **7**(14):2325–30.

26. Izawa, J., Rane, T., et al. (2008). Motor adaptation as a process of reoptimization. *J. Neurosci.* **28**(11):2883–91.

27. Kadiallah, A., Liaw, G., et al. (2011). Impedance control is selectively tuned to multiple directions of movement. *J. Neurophysiol.* **106**(5):2737–48.

28. Keysers, C. and Perrett, D.I. (2004). Demystifying social cognition: A Hebbian perspective. *Trends Cogn. Sci.* **8**(11):501–7.

29. Kohler, E., Keysers, C., et al. (2002). Hearing sounds, understanding actions: Action representation in mirror neurons. *Science* **297**(5582):846–8.

30. Kuniyoshi, Y., Yorozu, Y., et al. (2003). *From visuo-motor self learning to early imitation - A neural architecture for humanoid learning.* In *International Conference on Robotics & Automation*, Taipei, Taiwan, IEEE.

31. Kushida, D., Nakamura, M., et al. (2001). Human direct teaching of industrial articulated robot arms based on force-free control. *Artif. Life Robot.* **5**(1):26–32.

32. Maravita, A. and Iriki, A. (2004). Tools for the body (schema). *Trends Cogn. Sci.* **8**(2):79-86.

33. Meltzoff, A.N. and Decety J. (2003). What imitation tells us about social cognition: A rapprochement between developmental psychology and cognitive neuroscience. *Philos. Trans. R. Soc. Lond. B Biol. Sci.* **358**(1431):491–500.

34. Meltzoff, A.N. and Moore, M.K. Imitation of facial and manual gestures by human neonates. *Science* **198**(4312):75–8, 1977.

35. Moore, B. and Oztop, E. (2011). Robotic grasping and manipulation through human visuomotor learning. *Robot. Auton. Syst.* in press.

36. Moore, B. and Oztop, E. (2012). Robotic grasping and manipulation through human visuomotor learning. *Robot. Auton. Syst.* **60**(3):441–451.

37. Nehaniv, C.L. and Dautenhahn, K. (2002). *Imitation in Animals and Artifacts.* Cambridge, MA: MIT Press.

38. Obayashi, S., Suhara, T., et al. (2001). Functional brain mapping of monkey tool use. *Neuroimage* **14**(4):853–61.

39. Oztop, E. and Arbib, M.A. (2002). Schema design and implementation of the grasp-related mirror neuron system. *Biol. Cybern.* **87**(2):116–140.

40. Oztop, E., Chaminade, T., et al. (2005). *Imitation bootstrapping: Experiments on a robotic hand.* In *IEEE-RAS International Conference on Humanoid Robots*, Tsukuba, Japan.

41. Oztop, E., Kawato, M., et al. (2006). Mirror neurons and imitation: A computationally guided review. *Neural Netw.* **19**(3):254–71.

42. Oztop, E., Lin, L.-H., et al. (2006). *Dexterous skills transfer by extending human body schema to a robotic hand.* In *IEEE-RAS International Conference on Humanoid Robots*, Genoa, Italy.

43. Oztop, E., Lin, L.H., et al. (2007). *Extensive human training for robot skill synthesis: Validation on a robotic hand.* In *IEEE International Conference on Robotics and Automation*, Rom.

44. Oztop, E., Wolpert, D., et al. (2005). Mental state inference using visual control parameters. *Brain Res. Cogn. Brain Res.* **22**(2):129–51.

45. Rizzolatti, G., Fadiga, L., et al. (1996). Premotor cortex and the recognition of motor actions. *Cogn. Brain Res.* **3**(2):131–141.

46. Saunders, J., Nehaniv, C.L., et al. Teaching robots by moulding behavior and scaffolding the environment. In *Proceedings of the First ACM SIGCHI/SIGART Conference on Human-Robot Interaction*. Salt Lake City, Utah, New York: ACM: 118–125, 2006.

47. Saunders, J., Nehaniv, C.L., et al. (2007). Self-imitation and environmental scaffolding for robot teaching. *Int. j. Adva. Robot. syst.* **4**(1):109–124.

48. Schaal, S. (1999). Is imitation learning the route to humanoid robots? *Trends Cogn Sci.* **3**(6):233–242.

49. Schaal, S., Ijspeert, A., et al. (2003). Computational approaches to motor learning by imitation. *Philosophical Transaction of the Royal Society of London: Series B, Biological Sciences* **358, 1431**:537–547.

50. Shadmehr, R. and Wise, S.P. (2005). *Computational Neurobiology of Reaching and Pointing.* Cambridge, MA: MIT Press.

51. Steffen, J., Oztop, E., et al. (2010). *Structured unsupervised kernel regression for closed-loop motion control.* In *IEEE/RSJ International Conference on Intelligent Robots and Systems (IROS).*

52. Tani, J., Ito, M., et al. (2004). Self-organization of distributedly represented multiple behavior schemata in a mirror system: Reviews of robot experiments using RNNPB. *Neural Netw.* **17**(8-9):1273–89.

53. Umilta, M.A., Kohler, E., et al. (2001). I know what you are doing: A neurophysiological study. *Neuron* **31**(1):155–165.

Section II

Emulating the Neuro-Mechanisms with Humanoid Robots

3 Hands, Dexterity, and the Brain

Helge Ritter and Robert Haschke

CONTENTS

3.1 HUMAN DEXTERITY AND COGNITION

Our hands are centrally involved in many of our daily activities. Reaching for objects and grasping and manipulating them usually is an almost effortless activity. Whatever our hands do, it always appears very simple to us. Yet, as children we need many years to learn how to use our hands in increasingly sophisticated ways to feel, explore, grasp, and manipulate objects. Later on, we learn to use a large variety of tools to extend our manual capabilities even further and to connect them with various cognitive skills such as writing or the playing of musical instruments. Our hands are also important mediators of social contact: from early childhood, they are crucial to get into touch with and to feel others, to signify affection, and to enrich our communication with gestural expression.

This all is made possible by the seamless integration of our hands into our cognitive system, making our manual skills an important part of our interaction with the environment and of our capacities for feeling, exploring, acting, planning, and learning.

A deeper understanding of these skills might start from finding out the processes that enable touch and vision and how these modalities are then combined to achieve hand–eye coordination for grasping and manipulating objects. The sheer number of objects that we can grasp and handle makes the analysis of the involved processes a very daunting task, probably not any simpler than an understanding of language (the number of familiar words and of familiar objects perhaps being of the same order of magnitude). Moreover, handling of many objects is not just a matter of physics alone: when we open a bottle, we anticipate that we will access its contents by a very familiar sequence of further manual actions, possibly involving additional objects such as cups or spoons. Not only can we anticipate these actions, we also can imagine how the bottle, the cup, and the spoon will feel in our hands, and any deviation from our expectations triggers a rich repertoire of corrective actions so that we can finish most of our daily activities very safely, despite the absence of really precise information about the geometry of our food items, their friction constants, or their elasticity coefficients. This skill gives us the capacity to shape our environment in planned and coordinated ways to an extent not seen in any other species, exerting in turn a strong driving force for the evolution of a capability to envisage our actions before we actually carry them out. This may have prepared the final dissociation of physical and mental action, giving us the ability of "manipulating" imagined objects, of goal-directed planning, and, ultimately, of conscious thinking.

The so-far last step in this chain of developments seems to be communication, which is based on goal-directed re-arrangements of the thoughts in our conspecifics, appearing as the extension of the reach of our "mental hands" beyond our own thoughts.

It becomes obvious that any deep understanding of human dexterity will almost inevitably lead us into elucidating much of the essence of cognitive interaction from the "physical" sensorimotor level straight up to the highest levels of thinking, language, social, and even emotional interaction. And it becomes clear that any single "theory" can only contribute a highly partial view: the richness of human dexterity, manual action, and its embedding in cognition poses as its perhaps foremost challenge to understand an architecture of interwoven processes that together can bring about this astonishing phenomenon.

This chapter shows how robotics and the quest for a deeper understanding of human dexterity share many research questions, making it natural and productive to look for bridges between the disciplines in order to arrive at a comprehensive picture of what it requires to make hands versatile and central tools for a cognitive system.

In doing so, we do not discuss any of the robots that are in use on today's factory floors, where they insert windows into cars or perform similar nontrivial assembly steps in narrow domains and in highly repetitive manners. Our interest is on anthropomorphic robots whose body shape enables them to carry out movements that can be very similar to our own. In addition to their obvious potential as useful future assistants that are better matched to comply with our home environments with their devices, furniture, and architectural features all tailored to our human body structure, these robots offer a novel type of research tool to test and develop ideas of how embodied cognitive interaction can work and can be created.

The neurosciences, cognitive psychology, movement science, linguistics, and the social sciences all provide us with different perspectives and different levels of description of the processes that contribute to manual dexterity and its role in cognitive interaction. Although simulation can be a powerful first step to test and integrate some of these insights, simulation easily "falls victim" to idealizations and a necessarily simplified modeling of reality.

In contrast, physical robot platforms offer real-world tests. They can offer stringent proofs of the workability of ideas about how dexterity can arise and how it can be functionally interconnected with our remaining cognitive facilities. Thereby, robots can provide us with strong "idea filters," helping us very early to identify what we need to implement a particular skill in an interactive system that has to work with real sensors, in real-time and under the limited accuracy of an embodied physical system. For instance, they make us highly aware when a grasping algorithm only works with a precise geometry model of the to-be-grasped object, or when it requires accurate data about the friction between the fingers and the object. They confront us with properties that we often tend to idealize away in our models, such as elasticity or very "non-Cartesian" object shapes of food items. And they highlight what it takes to come closer to the superb "technology" of nature, with hands covered densely with tactile sensors, so far unmatched weight–force ratios, and the real-time control of fast and sophisticated movements with "wetware" that is exceedingly slow from an engineering perspective.

Moreover, endowing such systems with increasing manual competences can suggest new experiments and open new windows into the processes underlying dextrous manipulation. For instance, creating better tactile sensing allows us to observe the touch patterns accompanying human hand actions. Designing different anthropomorphic hand shapes provides us with insights into the role of geometry for hand dexterity. Enabling robots to cope with a multitude of objects and actions challenges our understanding of how such skills need to be represented and how representations in different modalities can get coordinated. Finally, bringing such robots to the point of cooperating with humans can help to connect insights from the neurosciences about imitation learning with robotics research into rapid learning mechanisms and findings from the social sciences about human–robot cooperation.

In the following, we first consider the goal of creating multifingered manipulators whose capabilities can approximate human dexterity to some degree. We highlight some of the major issues involved and describe some representative state-of-the-art hand systems and their properties. Next, we move our focus to the perhaps most fundamental task for a hand: the grasping of objects. We compare different computational and neuroscientific accounts for the necessary processes, ranging from hand–eye coordination to frameworks for grasp characterization, hand preshape selection, and different approaches for realizing grasps. As a next step toward higher-level skills, we discuss the challenges that are associated with the controlled manipulation of a grasped object. This will leads us into a discussion of how objects and actions can become represented for manual action, and how manipulation of objects gives rise to further questions, such as how to handle deformable objects.

We then focus on a different aspect, namely the use of the hand as a perceptual device. This also gives us the opportunity to contrast properties of the human hand with current technological approaches.

As a last major aspect, we briefly summarize some major findings and ideas about the role of the hand and brain for communication. Finally we discuss the significance of a better understanding of "manual intelligence" for a deeper understanding of cognition as a whole.

We are aware that our discussion inevitably is far from exhaustive. Given the limited amount of space, we had to leave out many important aspects of the field and to focus on characteristic examples to illustrate major ideas and approaches. Many of our choices are undoubtedly affected by a strong bias from robotics in general, and from our own research interests in particular, but we have attempted to bring out at least some of the numerous connections among robotics, neuroscience, and cognitive psychology that make up much of the fascination of the field.

3.2   DEXTROUS ROBOT HANDS

Loss or injury of limbs in warfare was since ancient times a major motivation for attempts to replicate arms and hands in order to provide useful or at least cosmetic prostheses. Over centuries, these constructions were rather crude, such as the famous hands of the German mercenary Götz von Berlichingen in the sixteenth century, which had to be actuated by an arrangement of catches and springs.

The sophisticated function of the human hand with its more than 20 independent degrees-of-freedom (DOF), actuated by more than 30 muscles and aided by numerous proprioceptive and tactile sensors, remained for a long time far beyond the reach of any human technology. This began to change only in the late twentieth century, when advances in materials, actuators, sensors, and control electronics offered possibilities for more realistic approximation of our most dextrous extremity. Digital computers and robots created additional interest in versatile robot manipulators and the obvious challenge to realize anthropomorpic robot hands in order to bring dexterity to robots. A timeline is shown in Figure 3.1.

The Belgrade hand and the Utah–MIT hand belonged to the first hands designed with that goal in mind. They had an anthropomorphic structure with an opposable thumb and their design anticipated some later perfected features, such as coupled joints or the use of tendons to realize the control of a large number of densely arranged joints. These systems were influential in the sense that they contributed to an early appreciation of the factors that need to be considered for realizing hands that can approximate human dexterity.

A first and very fundamental factor is the arrangement, size, and moveability of the fingers. Analyzing the evolution of hands and hand use in primates [72] has revealed that a seemingly simple change, the specialization of one finger as a highly movable

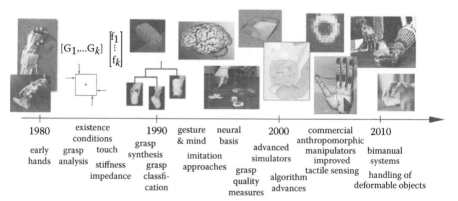

FIGURE 3.1   Bringing dexterity to robots and elucidating its underpinnings. Coarse timeline of major research topics and developments since the availability of the first articulated robot hands.

"thumb" acting in opposition to the remaining fingers, brought a dramatic step for the evolution of human dexterity, whereas the precise length ratios of the fingers seemed to be much less critical, as indicated by the significant variability of human finger-length ratios.

Simulation packages, such as the GraspIt system [42] or the more recent Open-Grasp simulator [8] have become very effective tools for studying the impact of hand kinematics on the attainable dexterity. These systems allow us to simulate how the given hand kinematics constrains the attainable contact patterns between the hand and arbitrary rigid objects. In this way, they allow accurate predictions about the grasps of which a particular hand design will be capable. Yet, kinematics is only one among many factors contributing to a versatile hand.

Actuators pose an entirely different set of constraints. Human grip forces can reach 400 N and more [58]. Even current technology cannot provide sufficiently strong miniaturized actuators that would fit into the finger phalanxes or at least into the hand palm, a constraint apparently shared with nature and necessitating "extrinsic" actuators, placed in the forearm and using tendons to transmit forces to the finger joints.

This actuating principle had already been adopted in the above-mentioned "historic" hand systems and recurred in many of their successors. A modern example is the Robonaut2 [1] hand which has been designed for space operation and whose kinematics has been extensively optimized in simulation. A major innovation of this hand was the development of a novel tendon material ensuring very high break forces, low tendon friction, and high durability against abrasion. The tendons are actuated with DC motors positioned in an integrated forearm. The fully exerted fingers can exert a tip force of more than 20 N and a tip speed of 20 cm/s. The hand has 12 independently controllable DOF and is only moderately larger than a human hand. Like many other designs, this hand is "underactuated", that is, it possesses more joints than controllable degrees-of-freedom. These "surplus" joints are controlled by coupling them in a fixed pattern to the movements of the remaining, controlled joints. Although the most frequently adopted pattern is a fixed coupling of the movements in the two outermost finger joints, more sophisticated adaptive schemes may enable parsimonious designs for robot hands that can be very dextrous while requiring only a modest number of controllable degrees-of-freedom [24].

Recent breakthroughs in the miniaturization of electric motors have also created the possibility of hand designs with "intrinsic" actuators, however, at the expense of somewhat reduced finger force levels. A major milestone for such an extremely integrated hand has been the DLR-II hand with four fingers (one opposed as a "thumb") with integrated brushless DC motors [4]. Each finger has three DOFs, and position and torque sensors integrated in all joints enabled to control "programmable stiffness" for all degrees-of-freedom. The available motors at the time of the construction of this hand enforced an overall size significantly larger than a human hand; however, further advances in motor miniaturization, along with additional ideas and refinements of the design concepts, have enabled the construction of the successor model DLR-II-HIT [41] (see Figure 3.2) with 15 DOF and a size that is only moderately larger than a human hand, while still capable of finger forces of up to 10 N.

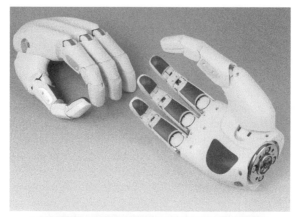

FIGURE 3.2   Two modern anthropomorphic robot hands. (Top) DLR HIT hand with integrated electric motors and 15 DOFs. (Bottom) tendon-driven Shadow Dextrous Hand with 24 DOFs (for details, cf. text).

If the primary use of a hand is for a prosthesis, weight becomes of paramount importance. Some recent prosthetic hand designs demonstrate that reducing the design to a small number of very carefully selected degrees-of-freedom to sacrifice on actuator weight, and using light weight plastic materials instead of metal led to light hands that still can carry out a useful number of different grasp patterns. Such hands typically have only a single DOF per finger and may delegate the operation of some DOFs even to the other (healthy) hand of the wearer, for example, for switching between sideways or opposition position of the thumb [60].

Hydraulic or pneumatically driven actuators offer an alternative to electric motors that can deliver high forces with low weight. The FRH-4 hand developed at KIT [7] is a modern hand that realizes a high power-to-weight ratio through lightweight fluidic actuators directly integrated into the finger joints. The hand has 11 joints, grouped into eight independently controllable DOFs. A pressurized pneumatic medium inflates the actuator and thereby causes a rotary motion of the associated joint. Pressures are controlled using 16 digital valves (controlling in- and outflow of the medium for each DOF). The actuators can work both with a hydraulic medium or with pressurized air, which yields higher compliance at the expense of a more nonlinear behavior. This offers additional challenges for the realization of suitable force-position control schemes, but simplifies the hand design because air can simply be released into the environment. The actuator torques are always in the same direction; an antagonistic rubber band provides the required retraction force. Joint positions are measured in 12-bit resolution through magnetic Hall-effect rotary sensors. Additional air pressure sensors allow improvement of the control scheme. Joint angle control is achieved through a cascaded control scheme for pressure and joint angle, with pressure in the inner loop.

Speed offers another challenge dimension. A significant part of our dexterity relies on the ability to make very fast finger movements. This allows us to catch thrown objects, to flip something between our fingers, to type rapidly, or to play musical instruments. The objective for the high-speed hand reported in Reference [49] has been to create a research platform for these dynamic aspects of manual action. This requires high speeds and accelerations and, therefore, low weight of the movable parts. To fulfill these requirements and realize a high degree of dexterity at the same time, the hand has three identical fingers, each with two joints. The two outer fingers have an additional joint allowing their movement in opposition to the middle finger. Each of the total of eight joints is actuated by an integrated high-speed motor specially designed to tolerate short, very high bursts of input current. As a result, each joint can accelerate within 10 ms to its maximal speed of 1,800/s. Using this hand in conjunction with a specially developed high-speed vision system, the authors have been able to demonstrate active catching of free-falling objects, or the very impressive dynamic regrasping of a bricklike object, using a strategy of brief throwing and recatching.

A high number of suitably arranged degrees-of-freedom for finger movements is perhaps the most crucial prerequisite for achieving good dexterity. One of the most leading designs in this regard is the Shadow Dextrous Hand [13], which is also one of the very few highly dextrous robot hands that are commercially available (see Figure 3.3). It is human-sized with 24 degrees-of-freedom, 20 of which are independently controllable. In addition to two independent DOFs for bending, each finger can independently be turned in the lateral direction. Together with a 5-DOF thumb this provides the necessary prerequisites for in-hand object manipulation. An extra degree-of-freedom in the palm aids the execution of power grasps. The hand exists in two versions, using for each joint either a McKibben pneumatic "muscle" actuator that contracts under the inflow of pressurized air, or an electric DC motor to pull the tendon. The motor version is the more recent one and also contains design improvements in the tendon routing and pulley attachments, leading to considerably smoother movements as required for dextrous action.

Accurate finger control requires sufficient kinesthetic feedback to compensate model inaccuracies. Different sensing devices, such as potentiometers, Hall-effect or

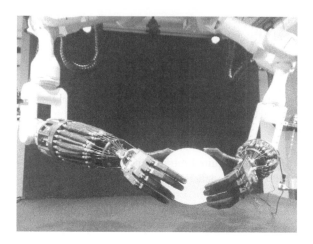

FIGURE 3.3   Bimanual research system with a pair of anthropomorphic robot hands (Shadow Dextrous Hand) mounted on robot arms (PA-10) for positioning. Each hand has 20 and each arm has 7 independently controllable degrees-of-freedom (DOFs), resulting in a 54-DOF platform. Cameras and tactile sensors provide perception.

optical sensors, have been developed to provide accurate information about finger joint angles and joint forces. Even more demanding is the realization of good "cutaneous" sensing through tactile sensors on the finger and hand surface. If each finger has only a single contact in known position, the contact force is computable from the measured joint torques. In all other cases, only a more-or-less extensive coverage of the finger segments and the hand palm with tactile sensing elements can provide detailed information about the forces exerted by a grasp or during manipulation. A wide range of sensing principles has been explored to realize different approximations to the so-far unattained tactile sensing of the human hand (see Section 3.5 for a discussion of the hand as a perceptual device and references therein).

Finally, the contact properties of the fingertips and the palm are major factors for the attainable grasps. Human fingertips are soft and offer good friction on a wide variety of object surfaces. Analyzing the physics of such finger contacts is a complex problem [45] and most analyses and optimizations of robot grasps are based on contact models that are simpler to deal with, but are less favorable for good grasps.

Most of our daily actions are bimanual. To explore and synthesize such skills for robots requires sophisticated research platforms that allow us to bring two robot hands in close opposition to create a workspace with a similar geometry as in human manual action. In addition to dextrous hands this requires highly movable robot arms, preferably with more than 6 DOF per arm to provide the system with redundant degrees-of-freedom for flexible positioning of its hands in a variety of interaction situations. A typical platform is depicted in Figure 3.3. It features two Shadow Dextrous Hands mounted on Mitsubishi PA-10 arms, each of which provides 7 DOF to facilitate avoidance of singularities in the workspace. The entire arm–hand system comprises a total of 54 independent degrees-of-freedom. There are 24 Hall sensors per hand to provide accurate joint angle feedback to control the 80 miniature solenoid on–off valves that adjust air in- and outflow into the pneumatically driven "muscle"-like actuators transmitting their forces via tendons to the fingers. The system is complemented with a Kinnect camera for 3D object segmentation and monitoring of the workspace [70].

Despite still being far away from the capabilities of human hands, platforms such as these begin to cross the critical threshold beyond which one can begin to study issues of advanced manual action in a robotics setting.

3.3   GRASPING

Being able to grasp objects allows us to adapt our environment instead of adapting ourselves. Grasping is also the major activity that "brings us into touch" with the world around us. Grasping allows us to feel what we just have seen, thereby constantly connecting our visual and our tactile modality. And, very importantly, grasping leads us from passive perception to active control.

Considering the seemingly simple act of grasping an apple, we notice that it begins with a shift of our visual attention that prepares a coordinated action of motor commands to the hand and to our eye when reaching for the object. This action is accompanied by sensorimotor and visual feedback to control the hand such that it safely approaches the objects in proper orientation and preshape. It proceeds with a coordinated closing of the fingers until their rich sensors signal us a familiar haptic pattern that finally confirms that we have made physical contact with the expected object and in the expected manner for the apple now to be fully at our disposal. And any significant deviation from its expected progression swiftly initiates corrective actions toward ensuring the intended outcome.

When we grasp the apple, all of this complexity is hidden from our conscious perception, making us entirely unaware that our brain has just performed a highly amazing act. An ultimate understanding of how this became possible might be sought in the underlying neural processes, scattered across the brain areas that were involved in the action. The combination of modern imaging methods and numerous painstaking single-cell studies has revealed a substantial number of brain areas connected into what has been termed a "grasping network" [17]. The major inputs to these circuits stem from the visual and the somatosensory systems, whose operation is at best understood at their lower levels, but much less with regard to the higher processed outputs that are sent to the grasping circuits. This makes it very difficult to elucidate how the different parts of the grasping network work together. In addition, most studies are limited to primates, whose dexterity is considerable but yet in stark contrast to what human dexterity can do [9]. Therefore, complementation of neuroscience experiments with computational modeling, the use of computer simulations, and experiments with real robot hands can be an invaluable source of additional information for analyzing different hypotheses about grasp action control.

Such a wider approach also brings into view additional layers of description and understanding: in addition to a "microscopic theory" at the level of the constitutive neural processes, we may strive for "coarser grained" theories focusing at the physical and control level, at the level of behavior, or even at a cognitive level, concerned with internal goals and meaning. Some of these levels may be more easily accessible than others, thus helping to find entry points from which our understanding then might work forward to those levels that are very hard to access directly.

The best observable layer of dextrous grasping is perhaps the hand–arm kinematics itself. Focusing on the parameter of grip aperture, Jeannerod in his now classic work [32] observed a highly stereotypical pattern of gradually increasing grip aperture to a maximum value that is highly correlated with object size and is reached near the last third of the movement, after which it shrinks again until the fingers make contact with the object. This pattern has been confirmed in many subsequent studies that extended the investigation to dependencies on object properties beyond size and on further kinematics parameters, such as object distance. (For a review, see Reference [3].)

Grip force is another important parameter. The proper assignment of grip forces constitutes a difficult task, inasmuch as it depends not only on the shape of the object (which can be seen), but also on parameters such as weight and friction properties, which can only be inferred indirectly and with considerable uncertainty. Moreover, grip forces may need to be rapidly adjusted to prevent slippage of the object as a result of disturbances, or when the object shall be accelerated, such as during lifting. Studies of grip force adjustment [36] have revealed that the brain manages to assign the required grip forces in a very parsimonious manner that keeps only a small safety margin against slippage while coping with a wide range of different situations. An important element in this strategy is a rapid grip tightening reflex that is triggered by cutaneous sensors that react to the vibrations that accompany slippage.

But grasps are also affected by factors that do not exist at the level of physics and geometry alone but are rooted in the anticipation of effects that the grasp will have in the future. A well-studied example is the optimization of "end-state comfort" [57]: when an action, such as putting a mug into a dishwasher, requires releasing the object in a "reversed" (upside down) orientation, we tend choose the "awkward" reversed orientation of the hand for the starting grasp of the object, so that releasing the object in reversed orientation makes the arm and hand end up in their comfortable, normal state. Interestingly, this ability is shared with primates, but it is absent in very young children, who only develop it after several years of grasping practice. Practice lets us also anticipate numerous even more subtle constraints for choosing a grasp, for example, when grasping food items, objects with dangerous or fragile parts, or during handover from another person [73]. In all these cases, our grasp choices are informed by rich background knowledge about the object and a nontrivial understanding of the purpose of the grasp.

The most ambitious approaches are aiming at a replication of the architecture of the "neural grasping network" on robot systems [19,37,50]. Given that our knowledge about these networks still is very limited, these attempts have to fill many gaps with tentative assumptions. On the other hand, this allows us to compare the impact of different assumptions and may generate helpful feedback to refine the focus of neuroscientific studies. One guiding idea is to conceptualize grasping as essentially a multistage mapping problem: the visual system extracts an initial representation which then is further transformed along separate mapping pathways into object location and into grasp-relevant object features, such as shape, size, and orientation.

Next, these features are associated with potential grasp types that are suitable for the object and task. From this set of candidates finally a suitable grasp type is chosen and mapped to suitable finger and arm parameters. This requires the connection of object location and object shape as well as orientation information, because suitable arm–hand configurations are affected by each of these factors.

An early concretization of this general idea is the FARS model [19]. It suggests that the first mapping step into grasp-relevant object features occurs within area AIP in the macaque brain, where one finds neurons selectively tuned to grasp relevant object features such as size, shape, and orientation. The second stage, the association with potential grasp configurations and their context-dependent selection, is assumed to be carried out in an area denoted as F5. This area receives inputs from other brain areas, making it a good candidate to satisfy task- and context-dependent constraints when selecting a grasp that then would become executed by passing suitable information from F5 to the motor area M1 which, according to the model, represents the last step in the mapping sequence by creating the finger and arm motion commands for a coordinated arm–hand movement.

Although the FARS model leaves out many details (such as how grasps and contextual constraints are represented, and how these representations interact), its major merit is to provide an at least coarse computational framework for the overall task that paved the way for subsequent refinements, such as the addition of learning. One approach has considered using the selected grasps to derive feedback corrections to the output that was delivered by the previous mapping stages [50]. This model shares with the FARS model its formulation at a very high level of abstraction, leaving many options—and challenges—of how to implement the required steps in more concrete and realistic situations. An exemplary recent attempt in this regard is [51], where the authors have implemented the mapping cascade from the visual input to the hand shape output very concretely. It has a four-layered visual system of alternating "simple" and "complex" cell layers that extracts a set of visual features that are sufficiently correlated with invariant object properties, and is followed by a mixture model that maps the visual features into a probability density on the space of grasp shapes. A final selection stage uses a heuristic scheme to select the most promising "probability peak" that identifies the grasp posture finally applied to the object. This model goes very far in filling in very plausible neural models for the required transformation steps; however, it still is restricted to a simulation study.

A complementary line of modeling leaves out the issue of how the required mapping steps might be mapped to neural structures and focuses on the computational analysis of essential building blocks for the overall process.

An obvious and conceptually appealing starting point is the physics of the grasping situation, shown in Figure 3.4. In the simplest idealization, all contacts are modeled as point contacts with friction, imparting forces and torques on the object. The resulting situation can then be analyzed with respect to the net forces and torques, and, more importantly, with regard to the stability of the grasp, for example, characterized by the ability to resist external forces and torques. A first characterization is through the concept of "force closure" which requires a grasp to be able to resist arbitrarily directed, but small forces and torques on the object by making small modifications to the grasp forces. An extension of force closure [21] considers more realistic finite

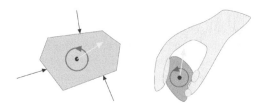

FIGURE 3.4   (Left) Idealized grasping. A detailed geometrical model is used to compute optimized grasp points and forces based on explicit contact models, such as point contacts with friction. (Right) Real-world grasping. Hand preshape and topological "encaging" leads to robust grasp despite the lack of detailed object geometry and lack of precise knowledge of contact conditions.

disturbances and characterizes the "volume" (the "grasp polytope" in the 6D product space of forces and torques) of disturbances that can be resisted by the grasp under given bounds of the available contact forces. Summarizing the 6D volume by the radius of the largest inscribable sphere leads to a compact grasp quality measure and, using ideas from semidefinite programming, has enabled the formulation of systematic optimization algorithms for grasps w.r.t this or similar grasp quality measures [29,30]. This can be computationally costly, but Borst, Fischer, and Hirzinger [12] show that simple randomized generate-and-test approaches can provide good approximations to theoretically optimal grasps, even for hands of the complexity of the DLR-II hand. These optimizations are possible with precisely known geometry and friction models for the involved object, the manipulator, and their mutual contacts.

   In real-world situations, such information may be unavailable or very difficult to obtain. This has motivated approaches that de-emphasize the need for a detailed, physics-based modeling of the interaction situation at the contact points. Instead, they take a more "topological" stance, either trying to identify suitable contact point patterns from visual features, or to abandon the idea of specifying contact points entirely and instead consider a grasp as a dynamic process that seeks its own stable contact points as a result of an "attractor dynamics" arising from a suitably prescribed finger motion started from suitable inital conditions ("pregrasp").

   Examples from the first category of works are found in References [5,62]. Instead of obtaining grasp point candidates from a mathematical optimization scheme, the researchers use a trainable classifier to assign grasp point locations based on visual input of images of objects. Starting from a collection of training objects with human-marked grasp point locations, they create a large dataset of computer-rendered images containing these training objects in different poses and contexts to achieve generalization. Subsequently, they use this enlarged training set to train their classifier in a supervised fashion to generate grasp point candidates for novel objects that are similar to the training instances. Thereby, they demonstrate how human grasp point selection knowledge can become "compiled" into a vision front end that can suggest grasp point candidates for a robot hand, using visual scene (stereo) images as its only input. An encouraging result is that the trained system can generalize rather well both with regard to the positions of the training objects as well as with regard to novel, similar real-world objects, although thus far the method has been demonstrated only for grasps with "opposing" grip point locations that have been realized by simple two- or three-finger grippers.

The second category of approaches is motivated from observations of human hand–eye coordination that have revealed that many human grasps pass through a "preshape" phase [32], in which the hand is already close to but not yet in contact with the object and shaped such that the fingers begin to surround the object in a cage-like manner. In the final phase the fingers close, thereby shrinking the cage and bringing the object into a stable position within the hand (see Figure 3.5). This appears to be a very attractive "holistic" strategy that requires only a minimum of detailed information, because many details of the final grasp configuration "emerge" as a result of the interaction between fingers and object during the shrinking phase of the cage.

The choice of a proper pregrasp can take inspiration from prior work on human grasp taxonomies [16]. This has revealed that the grasps that humans use in different situations can be organized into a taxonomy tree with only a small number of major branches that include power grasps in which the hand wraps around the object to create a large contact surface in order to be able to impart large forces, two- and three-finger ("tripod") precision grasps, in which object contact is restricted to the fingertips in order to maximize controllability of the object, and grasps that are intermediate between the precision and power grasps, using the thumb in opposition to all other fingers to hold the object. (For a more recent review, see also [58].)

These pregrasps can be directly mapped to anthropomorphic robot manipulators. Using these grasps Röthling et al. [56] have implemented the "preshape" approach for different robot hands: a robot hand with only three fingers to form the cage, and the Shadow Hand with five fingers and 20 independently controllable DOFs (see Figure 3.6). The five-fingered hand produced superior results; however, the cage-based strategy is even implementable for a gripper with only three fingers, although the range of graspable object shapes then is somewhat reduced.

The main grasp-related features required for this method are the preshape of the finger cage and its positioning relative to the target object. Experiments with the Shadow hand revealed that four different pregrasp shapes are sufficient to grasp the majority of a set of 20 typical household objects. Therefore, for many situations the task of determining precise grasp point locations can be reduced to just a discrete choice among a small number of hand preshapes, combined with a suitable finger closing process whose dynamics generates the details of the grasp. As a result, the visual mapping task can be significantly simplified to a categorization of the object into a small set of preshape categories plus a rejection class for those object shapes that may require a more sophisticated grasp.

An interesting element of this way of grasping is that an essential part of the computation becomes "moved into the interaction physics" between the hand and the object. This demonstrates how embodiment can simplify the control of actions. It seems that grasping has a very strong embodiment component where substantial benefits arise from the softness of the finger and hand surface and their excellent friction properties. These factors can significantly facilitate contact formation and grasping. Although algorithms that attempt to model and exploit these factors in detail

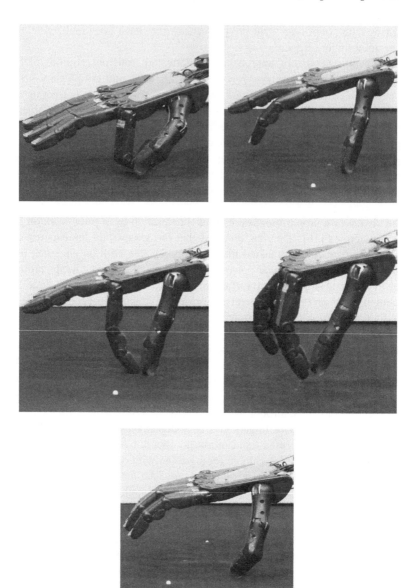

FIGURE 3.5   Hand preshapes for enabling robust grasps with the Shadow Dextrous Hand.

are very difficult to formulate, it may be much easier to formulate computationally lightweight strategies in which these effects are exploited without modeling them explicitly, and in such a manner that the remaining computations are simple and robust. Hence, an overarching lesson might be that we should be wary of the construction of "fragile algorithmic clockworks," needing information about a lot of details which are hard to know in real-world scenarios (such as friction, elasticity, precise shape, and

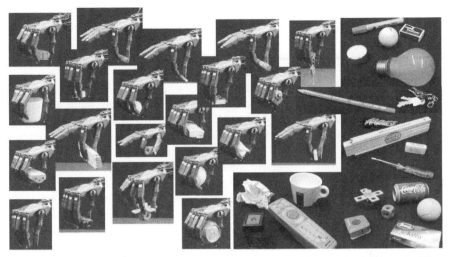

FIGURE 3.6 Preshape-based grasping of daily objects with Shadow Hand. (From Röthling et al. 2007. With permission.) See color insert.

mass distribution) and look instead for "holistic solutions" in which robust dynamics develops attractors toward desired states without creating too much need of making underlying model parameters explicit.

An interesting further line of developments combines elements from the previous two categories of approaches [11,43]. Given the availability of 3D geometric models for an increasing number of everyday objects, these authors take a data-driven approach and take a number of objects for which associated optimized grasps are determined by some method (such as algorithmic computation or prespecification by the human). These grasps then include both a specification of contact points and a specification of a suitable hand orientation and shape w.r.t. the grasped object. These associations, together with robustly determinable visual object features, are stored in a large database and the task of the vision system is to extract image features that are suited to index reliably into this database. At first sight, this appears to be a brute force method, but it allows for several optimizations, such as shape decomposition techniques to enhance the generalization ability from stored to novel object shapes. Associating grasps with salient object parts instead of entire objects can reduce the required storage to a much smaller database of "predictive" object parts and allow the obtaining of good grasps for large sets of daily objects [55] when accurate geometry data are available, for example, from 3D vision.

3.4   MANUAL ACTION: FROM GRASPING TO MANIPULATION

Once an object has been grasped, our fingers enable us to manipulate it in many different ways. This is very different from most simple grippers, whose constrained degrees-of-freedom severely restrict the local motions that can be imparted to the grasped object. At the same time, manipulation planning and control for anthropomorphic hands pose sensorimotor challenges that are even greater than those associated with grasping alone and that may even have acted as a major evolutionary drive for the evolution of higher cognitive abilities, such as tool use and language.

A thorough understanding of multifingered object manipulation has also become of significant interest in robotics because the advent of advanced anthropomorphic hand designs allows us not only to test computational ideas about multifingered manipulation beyond simulation, but makes any results in this field of practical interest for future robots.

From a computational perspective, manipulation has some resemblance to walking: in both cases, the system must coordinate a sequence of contact patterns for achieving desired forces between the agent and an external object. This has given rise to the concept of "finger gaits," and initial theoretical analyses of manipulation have focused on the precise characterization of the conditions under which such gait sequences exist [34]. This, however, does not yet solve the problem of how such movements can be planned in a robust manner.

Although there exist numerous planning and optimization algorithms for movements that are "smooth," typical manipulation sequences are of a hybrid nature, connecting smooth state changes during which the hand configuration changes continuously (while maintaining its current contact pattern with the object) with discontinuous transitions that occur when the contact pattern changes. This happens whenever one or more fingers are lifted or set down to create a new contact.

This has given rise to the concept of "stratified state spaces" whose "strata" consist of subsets of hand–object configurations that share the same contact pattern [22]. Different strata are connected by discontinuous transititions between their associated contact patterns. Using very elegant concepts from Lie theory, the authors have developed a general method for planning movement sequences across strata in such spaces.

Finite state automata (FSA) offer an alternative for planning in stratified state spaces. The idea is to consider each stratum as a separate node of a graph in which strata with possible transitions are linked by arcs. The resulting graph (and the corresponding nodes of the FSA) then reflect the "coarse-grained" structure of the manipulation space when only contact patterns are distinguished. Each FSA node then is responsible for handling the trajectory piece within its stratum [67]. Such a scheme is very much in line with emerging ideas about how the brain may use tactile feedback and vision to enable dextrous manipulation sequences by properly sequencing "action phase controllers" in response to tactile events [20,33]. Moreover, the scheme is easily extensible to accommodate more levels, thereby offering a computational framework that can be related to recent ideas about the larger scale architecture of manual action. This may offer another example of how insights from neuroscience and psychology on the one side, and robotics research, can mutually stimulate each other.

A major limitation of the work in Reference [67] (and of similar approaches) is, however, its reliance on accurate geometry models of the objects. So far, this has restricted its use to simulation approaches.

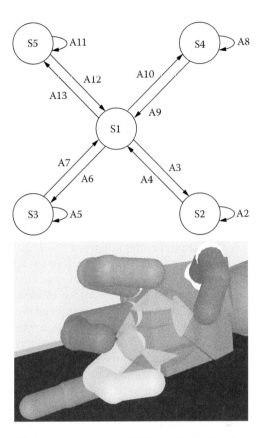

FIGURE 3.7   (Top) Finite state automaton (FSA) distinguishing five contact states S1–S5 connected by finger actions A1–A12. Each finger action corresponds to the activation of a specialized, low-level finger controller. (Bottom) Physics-based simulation of manipulating a sphere with the FSA on the left. See text for details. (Adapted from Li et al. 2012.)

In the following we describe a variant of the scheme proposed recently [Li+MHRB: 2012] and that, although thus far also limited to simulation studies, can lift the need for an accurate object model by combining FSM-guided manipulation with a local feedback scheme for finger control and a merging of finger-repositioning with an online exploration of neighboring object points.

Using four fingers and considering only states with three or four finger contacts, the possible transitions between state-space strata lead to a FSM of the structure depicted in Figure 3.7. Transitions between the center node and four surrounding nodes in this FSM correspond to changing the role of one finger from "support" to "local exploration" or vice versa. In the center node S1, a desired movement of the object is effected by suitable finger motions, assuming point contacts of all four fingers without slippage and rolling. The only approximate honoring of these assumptions causes deviations from the expected motions that are detected and corrected by sensing the true object motion through (simulated) visual feedback. Each of the remaining periphery nodes S2–S5 corresponds to one finger being lifted for local exploration. In these states, the object can be rotated through the remaining three fingers that are still in contact. Such actions do not change the contact pattern and are indicated by the "self-transition" arcs. The local movement of the exploratory finger is controlled under the influence of a measure that attempts to optimize a compromise between good object manipulability and good grasp stability when the finger regains its status of a support finger. A physics-based simulation is used to verify that the necessary steps can be carried out with information that is locally available at the fingers, plus accurate feedback about the resulting global object motion, when the object geometry is not "too intricate."

Some of the difficulties of taking the step from such simulations to real systems are connected with control challenges arising from the presence of kinematically closed actuator chains. This is a typical situation for manipulation, and when the actuators are rigid, small positional deviations can give rise to huge forces. For stiff actuators, this then requires very short control cycles of the order of 1 ms or below.

Biological systems cannot regulate motions at such timescales and use compliant structures whose inherent elasticity allows their control at much slower timescales. This has inspired robotics likewise to integrate elastic elements in actuator design, even if this may make models of such actuators more difficult to calibrate and analyze. On the positive side, the inherent compliance of elastic structures can be seen as a low-level control law that automatically provides corrective forces when the actual finger positions begin to deviate from the target configuration. This makes the system tolerant against small errors, for example, when carrying out an otherwise rigidly guided movement, such as turning a handle or unscrewing the lid from a jar. It also facilitates synthesizing manipulation sequences from approximate trajectory information, as obtained, for example, through observation of human manipulation trajectories.

Even then it is important to reduce the a priori very high state-space dimensionality of anthropomorphic manipulators. Modeling the hand as an only 12-dimensional manipulator (the available number of DOFs is actually significantly larger), and resolving only 3 different positions for each joint, would lead into a state-space of $3^{12} \approx 1$ billion different configurations, which would be impossible to visit in the duration of a lifetime (≈ 1.2 billion seconds). Therefore, actually occurring hand configurations must be highly correlated and "cluster" strongly in lower-dimensional "manipulation manifolds." For instance, Santello, Flanders, and Soechting [64] found that an only two-dimensional linear subspace can already capture 80% of the variability of a large number of natural hand postures that were recorded with a dataglove. This has encouraged studies that use techniques familiar from principal component analysis for projecting hand postures to lower dimensional spaces that are spanned by high-variance directions termed "eigengrasps" [10]. These linear methods can be generalized to employ manifolds, which, by virtue of their ability to "curve" in a nonlinear way, can capture even more variability with the same low number of dimensions.

Following this idea, Steffen, Haschke, and Ritter [35] generalized methods based on self-organized feature maps to create from raw human action capture data low-dimensional manipulation manifolds in which a highly structured manipulation sequence, such as uncrewing a cap from a bottle or jar, is represented as a motion of only a few major control variables [63].

In that case, the manipulation manifold can be as low as two-dimensional (one dimension for the cap radius, and one dimension as "progression time"), allowing the representation of segments of the cap-turning manipulation movement by smooth curves in this manifold. Using an inherently compliant manipulator system, in this case the 20-DOFs Shadow hand, allowed the demonstration of the opening of a marmalade jar as part of a complex, bimanual manipulation action [63].

Although the key point of this demonstration was on the finger action sequence for the opening of the jar, it also provides an example of an integration of several of the previously discussed aspects of manual interaction in a real-time operating robot system: a vision system recognizes the object location and posture to prepare target reaching of the first arm and proper hand–eye coordination for the initial grasp. A regrasp by the second hand then creates the required configuration from which the unscrewing motion can start. Finally, the action is concluded by lifting the cap once enough turns have occurred.

The actual implementation of apparently "simple" tasks such as this makes explicit that they are based on the coordinated activity of a substantial number of more elemental, constituent processes each of which itself can already be of substantial complexity. Without doubt, our robot solutions are still simple compared to the sophistication of the grasping networks that we begin to recognize in the brain. We may speculate that the detailed complexity of the biological circuits will forever remain unreached by our models, and system diagrams with a small number of "black boxes" may be too coarse to capture much of the essence of the underlying mechanisms. Actual robot implementations, therefore, may offer the chance of a useful "middle" level of abstraction, allowing us to "sketch" suitable processing architectures for manual action in a way that is sufficiently detailed to be validated with regard to their computational feasability and that may help to chart possible functional structures and thereby inform neurobiological modeling and interpretation of structures.

Defining four major levels of abstraction—sensorimotor control, sensorimotor representations, mental representations, and mental control—and assigning to them tentative computational structures that have been found useful to implement manual skills for robots as described here and in the previous sections, Maycock et al. [31,69] present a, still very tentative, attempt to connect ideas and concepts from robotics and psychology toward a general framework for manual action that bypasses the difficulties of matching processing abstractions with neural structures. Instead, this proposal focuses on the different axis of aspects that can be characterized functionally (Figure 3.8). Works such as this can also provide examples of the interdisciplinary research that is needed to make progress with the challenges of cognitive interaction.

In addition to such overarching contributions, there remains a strong need for focused advances for the better understanding of generic skills in manual interaction. In this regard, a particularly interesting challenge appears to be the handling of objects with "biological characteristics." These are objects that are soft and often deformable, such as plants, food, or fur. They tend to defy our usual representations that are

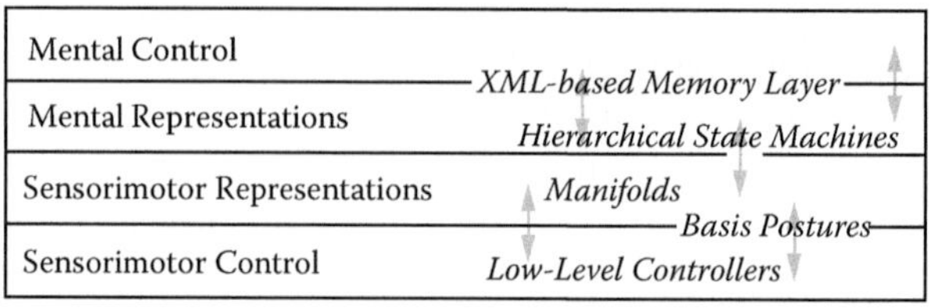

Mental Control		
	XML-based Memory Layer	
Mental Representations	Hierarchical State Machines	
Sensorimotor Representations	Manifolds	
	Basis Postures	
Sensorimotor Control	Low-Level Controllers	

FIGURE 3.8 Four-layer framework for manual action and possible implementation in robot systems: control layers at the sensorimotor and at an abstract "mental" level are connected through two representation layers that bridge the gap between these complementary levels of control. Experience from actual robot systems suggest implementation through indicated computational structures. (From Maycock et al., 2010. With permission.)

preoccupied with simple Cartesian shape primitives, such as cylinders, spheres, or polyhedra, and cause us instead to elaborate control models capable of reaching beyond the case of rigid or otherwise fixed structures. Manipulating such objects may also pave a way that connects insights about dextrous manipulation with a deeper understanding of the higher cognition without which the handling of such objects would not exist.

To gain insights into the associated challenges, we have begun to explore the handling of paper through robot hands. Paper has an intermediate position between fully rigid and fully deformable objects, and it offers an interesting scope of nontrivial interactions. Already picking up a flat-lying piece of paper can offer an interesting challenge even for human hands, and operations such as bending, folding, tearing, or crumpling of paper pieces are building blocks for higher-order capabilities that we invoke when we put something into an envelope or a bag, or when we use paper to construct even new objects manually.

Due to the deformability of the paper, the robot requires real-time visual feedback to be able to adapt its finger motions suitably with the changing paper shape, for example, for being able to "bulge" the paper suitably with one hand to enable the second hand to pick it up with a precision grip. See Figure 3.9. The work in Elbrechter, Haschke, and Ritter [14] shows how such feedback can be obtained, utilizing a physics-based modeling of the behavior of the paper. For operations such as folding, visual feedback has to be combined with force sensing to ensure proper task execution. This can be achieved by customized low-level controllers that are suitably sequenced with the help of a finite-state automaton properly designed for the task [15]. Examples such as these show how some of the complex multimodal coordination patterns that are typical of most of our daily manual actions can be achieved with current robot hands.

Human hand motions integrate a substantial variety of similar and many even much harder manipulation primitives. The challenge is to replicate a representative "vocabulary" of such manual actions and to develop a deeper understanding of how to blend them together into the sophisticated actions that make our hands such special tools of our cognition.

FIGURE 3.9   Bimanual folding of a piece of paper, using visual and tactile feedback. The paper is printed with fiducial markers to simplify visual perception. The inscribed colored grid depicts the robot's current model of the perceived paper shape. (Courtesy of Elbrechter, Haschke, and Ritter, 2012. With permission.)

3.5 HAND AS PERCEPTUAL DEVICE

When our hands get into touch with an object, they not only signify to us mechanical contact, but simultaneously provide us with information about the object's texture and material properties such as friction and thermal conductivity. Manipulating the object briefly between our fingers, we readily get access to further information, such as the object's weight, firmness, and shape properties.

The sensory basis of these capabilities is a rich sensory equipment of the epidermis. It has been estimated that the hand surface is covered by about 17,000 sensors. Four major sensor types have been identified that differ along the axes of spatial and temporal resolution [33]. Analogous to the visual fovea, there is also a highly nonuniform sensor density on the hand. It is highest at the fingertips, leading to a spatial resolution down to 0.5 mm, whereas the resolution falls to about 5 mm at the back of the hand.

The sensor responses are also affected by their embedding in an elastic skin whose friction and deformation properties during object contact has a significant effect on the sensory responses of the above-mentioned sensor systems. A striking example [59] is the important role that is played by the shape of the fingerprints, which could be shown to act as a sensitivity enhancer for discrimination of fine surface texture during finger movements across a surface.

In contrast to vision, tactile sensing is highly dependent on the active shaping of the contact between hand and object. Haptics is the associated interplay of cutaneous and kinesthetic sensing, involving numerous different processes that contribute to the formation of a haptic percept [LedKla2009]. Correspondingly, there is a participation of neural areas, of which the somatosensory and the motor areas, positioned adjacent and densely interconnected, are closest to the sensorimotor periphery. In both areas, there exist topographic maps of the hand surface, in which neighboring sensors are connected in a topographic fashion to neighboring cortical cells. These maps are adaptive and have been found to reorganize (e.g., to rededicate unused regions after the loss of a finger) while maintaining their topographic structure. Considerable adaptation can also result from extensive training and is reflected, for instance, in the significantly enhanced spatial acuity of Braille-reading people in their "reading finger tip".

A replication of similar capabilities in robot hands is still a rather elusive goal. Although there is progress toward the development of more and more capable "skin" sensors and their calibration [44], there is still a large gap regarding the sensing capabilities and spatial resolution of the human hand [65].

From a computational perspective, some aspects of early tactile sensing may be amenable to processing tactile "images" by similar algorithms as used for feature extraction and classification in early computer vision. This approach is also encouraged by apparent similarities in the representation of shape in the somatosensory and visual pathways [75], along with similarities in the characterization of receptive field properties, such as Gabor-like response profiles. These findings encourage the use of vision methods, such as the decomposition of tactile images into principal components, to enable haptic pattern discrimination and recognition for robot manipulators [28].

However, a major complication as compared to vision arises from the considerably increased complexity of the geometry of the sensor: whereas the visual fovea has a fixed spherical shape, the shape of the hand surface during tactile exploration of an object is highly variable, and the tactile "images" in the different touch sensor channels are accompanied by the proprioceptive kinesthetic information about the hand shape during manipulation.

How these processes interact is currently only very little understood. (For a review, see [33].) Available computational models usually focus on a single channel, such as the discrimination of object shapes through kinesthetic (hand shape) information when the hand is closed around the object, the discrimination of objects through their surface textures when moving a tactile sensor ("finger") across the surface, or the classification of spatiotemporal sequences of sensor images in simplified sensor geometries, such as when moving a planar tactile sensor matrix actively around objects [28].

However, robotics experiments can help to elucidate which features may be particularly informative for object discrimination, and how useful information is spread across different feature sets. Looking into this question and comparing over 100 heuristically chosen tactile features as a basis for tactile object classification, Schöpfer et al. [66] present evidence for a rather distributed representation, making it infeasible to "concentrate" a significant share of the information by simple linear methods, such as PCA, in a small set of principal feature dimensions for a typical scenario in which tactile information is gathered with an actively moved tactile sensor.

Another area of research is the construction of object representations from haptic interaction. Klatzky et al. pioneered the idea of "exploratory procedures" [40], which aim to reveal specific properties of an object, such as its friction, texture, or stiffness through suitably tailored actions, such as scratching, squeezing, or poking. There is no universally agreed definition of specific exploratory procedures; however, examples of how this concept can be implemented in robotics have been presented by several authors [71,76].

One challenge that is associated with haptic representations is the integration of geometry information together with information about stiffness and the sensing of movable degrees-of-freedom. For instance, we are easily capable of identifying the rotatory axis of a door handle through haptic exploration. de Schutter et al. [18] show how one can solve this task from a computational perspective and suggest an adaptive controller using Kalman filter techniques to replicate a similar ability for a robotic manipulator. The even more demanding task of identifying the movable part structure of a composite object through active exploration has been considered in Katz and Brock [38]. In their approach the authors substitute vision for tactile information, creating an object model from a set of feature point clouds generated from a sequence of exploratory "pushes" to an articulated chain of segments connected by rotatory joints. A different approach, building a 3D shape representation of rigid objects from a number of tactile "images" resulting from self-generated grasps by a three-fingered gripper with tactile matrix sensors in its finger pads, is shown in Meier et al. [46]. The authors also demonstrate that the resulting 3D shape representations can be used to recognize an object and discriminate it from a number of competitors.

Because currently the miniaturization of touch sensors into highly articulated, anthropomorphic manipulators still poses a difficult technical challenge, a complementary approach is to instrument the to-be-grasped objects with tactile sensors in order to create new windows into haptic interaction during manual actions. Recent work in the author's group has led to the construction of an "iObject" [39] that can sense and wirelessly transmit such haptic patterns and, thereby, allow us to investigate haptic control strategies during a variety of manipulation tasks.

3.6   HAND, BRAIN, AND COMMUNICATION

Studies of the cognitive development in children have provided numerous observations pointing to a close linkage of manual gesture, language development, and communication [25]. Manual actions for showing and giving develop before pointing, and pointing has been found to predict acquisition of single words. Later on, manual gestures become combined with word use and a further differentiation of manual gestures into deictic, metaphoric, iconic, and "beat" gestures develops and becomes intertwined with more elaborated language structuring [2].

These observations have led to a view that manual action shares with language an extensive sensorimotor basis (in which also the mouth region plays a major role, exemplified, e.g., in the Babcock reflex, causing babies to open their mouth in response to touching their palm), which leads to a highly correlated development of both modalities, or, in exceptional circumstances, to the ability to develop sign language if the vocal modality cannot be used, and the coupling of gesture and language in situated communication has begun to connect social robotics and computer graphics to add natural expressivity to agents by endowing them with the ability to enhance language with natural-looking deictic, iconic, or metaphoric gesture [6,47].

The tight coupling between manual action and language [52] is also reflected in the close proximity of the involved brain regions, such as the hand and mouth regions in sensorimotor and motor cortex, and remarkable cross-effects, for example, changes in pregrasp aperture when reaching for the same object, but while listening to words for large or small objects [61].

Most of our knowledge about brain control of hand actions stems, however, from studies with monkeys, such as macaques, which share with us the ability of fine manipulation. A vast body of studies has led to the discovery of a complex network of brain areas involved in the recognition, preparation, execution, and learning of manual actions that is now rather well known in macaques [17], with additional findings in humans that point to the existence of homologue networks in the human brain [68].

The elucidation of these networks is closely connected with the discovery of the so-called *mirror neurons* [54]. Activity in these neurons is correlated with specific actions, however, irrespective of whether the action is carried out by the animal, or is only observed by it. Thus, these neurons seem to represent or "mirror" specific actions per se, which has given rise to their name and the concept of a *mirror system* providing a representation of actions that can be shared by perception and behavior control alike.

The discovery of such a system has been a major missing link to explain how neural structures might subserve the ability of primates and humans to learn by imitating others. This in turn has led to ideas how this link between observer and actor might provide a "bridge" from "doing" to "communicating" [53], thereby setting the basis for the evolution of language [23] and connecting it with a capacity for manual skills. These ideas have received further support from recent neuroscience findings, such as evidence for even more generalized "mirror-type" neurons that may be involved in the creation of more abstract concepts required to enable higher-level action semantics, such as the representation of action goals [74]. Such abilities would appear crucial for creating linguistic representations and for enabling what has been termed the *social brain*, that is, the ability to infer goals and intentions in observed actions of others.

This social dimension leads back to another important role of our hands: using touch and haptics to enrich communication with an important emotional dimension. Although this important role of hands is clear from our everyday experience, systematic research into this subject is still in a very infancy stage, for example, investigating the role of touch for social communication [48], or how haptic interfaces could enrich social media [26].

3.7 MANUAL INTELLIGENCE AS A ROSETTA STONE FOR COGNITION

We have only been able to cover some of the major capabilities of our hands: grasping, manipulating objects, using our hands as perceptual devices, and, finally, some aspects of their role in communication.

A shared element of these capabilities is interaction. There exists an abundance of works aiming to explain intelligent behavior as arising from capabilities of perception, category formation, and decision making; however, the study of manual dexterity and its replication in robots forcefully entrenches us into issues of intelligent control.

In robotics, control is a very familiar concept. However, it usually is encountered within highly prestructured settings, with predefined assignments of statespaces and control variables, leading to well-formatted optimization problems, such as finding feedback laws that minimize a tracking error under a given set of conditions.

We believe that manual action requires a broader understanding of intelligent control. As we have seen in the preceding chapters, the relationships between hands and objects can be extremely multifaceted and are not well amenable to a picture where a controller with predefined state, input, and output variables is embedded in a fixed interaction loop of a known structure. Instead, the flexible shaping of these relationships—and thereby of the structure of the interaction loop itself—is an essential characteristic of the task. Coping with this challenge goes well beyond what current control approaches can deliver: although much of classical control theory is focused on the characterization of controllability for given situations and the derivation of suitable control laws, an important part of manual action requires solving the question of how controllability can be created in the first place through suitable "attachments" of the fingers, and their sensors (along with the eyes) to the object and its environment.

This challenging question is not solvable by a narrow focus on "low-level" aspects alone: controllability needs to be achieved on many levels that range from small local movements of an object to more global actions, such as unscrewing a jar, over the achievement of high-level goals, such as filling a glass of juice, up to the highest level that is encountered in communication: the structured interaction with thoughts.

As we have seen from the preceding sections, manual actions are crucially involved in all these levels. Therefore, the quest for a better understanding of manual action, along with its engineering facet of how to synthesize dextrous manual action for robots, is highly likely to catalyze deeper insights into what is required for an intelligent agent to become able to shape and influence its environment in increasingly sophisticated and abstract ways, starting with a control of the physics in its immediate surround, advancing to the mastery of a variety of mechanical devices and tools, and ultimately paving the way for grasping mental objects that are only present through thinking and communication.

Thus, research on dexterity, manual action, and how it is realized by the brain, seems to be in a pivotal position for the deciphering of the principles of cognitive interaction. Therefore, it has been argued that it might play a similar role as played by the Rosetta Stone for the deciphering of ancient writing systems [27]. At the same time, it can provide a fascinating unifying topic for elucidating an important and rich part of our cognition: the manual intelligence that becomes evident from what our hands can do in their daily actions.

ACKNOWLEDGMENT

This work was supported through DFG Grant CoE 277: Cognitive Interaction Technology (CITEC).

REFERENCES

1. Bridgwater, L.B. et al. The robonaut 2 hand - designed to do work with tools. In *IEEE International Conference on Robotics and Automation (ICRA)*, pp. 3425–3430, 2012.
2. Bates, E. and Dick, F. Language, gesture and the developing brain. *Develop. Psychobiol.*, **40**:293–310, 2002.
3. Smeets J.B. and Brenner, E. A new view on grasping. *Motor Contr.*, **3**:237–271, 1999.
4. Borst, C., Fischer, M., Haidacher, S., Liu, H., and Hirzinger, G. Dlr hand ii: Experiments and experiences with an anthropomorphic hand. In *IEEE International Conference on Robotics and Automation (ICRA)*, pp. 702–707, 2003.
5. Bohg, J. and Kragic, D. Learning grasping points with shape context. *Robot. Autom.*, **4**:362–377, 2010.
6. Hartmann, B., Mancini, M., and Pelachaud, C. Implementing expressive gesture synthesis for embodied conversational agents. In *Gesture in Human-Computer Interaction and Simulation*, pp. 188–199, 2006.
7. Bierbaum, A., Schill, J., Asfour, T., and Dillmann, R. Force position control for a pneumatic anthropomorphic hand. In *Proceedings of the 9th IEE-RAS International Conference on Humanaoid Robotics*, pp. 21–27, 2009.
8. Leon, B., Ulbrich, S., Diankov, R., Puche, G., Przybylski, M., and Morales, A. Opengrasp: A toolkit for robot grasping simulation. *In Simulation*, Vol. 6472, pp. 109–120, 2010.
9. Castiello, U. The neuroscience of grasping. *Nature Rev. Neurosci.* **6**:726–736, 2005.
10. Ciocarlie, M., Goldfeder, C., and Allen, P. Dimensionality reduction for hand-independent dexterous robotic grasping. In *Proceedings of the IEEE/RSJ International Conference on Intelligent Robots and Systems San Diego*, Vol. 29, Oct–Nov 2, pp. 3270–3275, 2007.
11. Goldfeder C. and Allen P.K. Data-driven grasping. *Auton. Robots* **31**(1):1–20, 2011.
12. Borst, C., Fischer, M., and Hirzinger, G. Grasping the dice by dicing the grasp. In *IEEE/RSJ International Conference on Intelligent Robots and Systems (IROS)*, Vol 4, pp. 3692–3697, 2003.
13. Shadow Robot Company. *Website www.shadowrobot.com/products.shtml*, 2012.
14. Elbrechter C., Haschke R., and Ritter, H. Bi-manual robotic paper manipulation based on real-time marker tracking and physical modelling. In *IEEE/RSJ International Conference on Robots and Systems (IROS)*, pp. 1427–1432, 2011.
15. Elbrechter, C., Haschke, R., and Ritter, H. Folding paper with anthropomorphic robot hands using real-time physics-based modeling. In *Proceedings of the IEEE International Conference on Humanoid Robots (Humanoids)*, pp. 210–215, 2012.
16. Cutkosky, M.R. On grasp choice, grasp models, and the design of hands for manufacturing tasks. *IEEE Trans. Robot. Autom.* **5**:269–279, 1989.
17. Davare, M., Kraskov, A., Rothwell, J. C., and Lemon, R. N. Interactions between areas of the cortical grasping network. *Curr. Opin. Neurobiol. 2011*, **21**:564–570, 2011.
18. de Schutter, J., Bruyninckx, H., Dutr, S., de Geeter, J., Katupitiya, J., Demey, S., and Lebfevre, T. Estimating first-order geometric parameters and monitoring contact transitions during force-controlled compliant motion. *Int. J. Robot. Res.* **18**:1161–1184, 1999.
19. Fagg, A.H. and Arbib, M.A. Modeling parietal-premotor interactions in primate control of grasping. *Neural Netw.* 11:1277–1303, 1998.
20. Flanagan, J.R., Bowman, M.C., and Johansson, R.S. Control strategies in object manipulation tasks. *Curr. Opin. Neurobiol.*, **16**:650–659, 2006.
21. Ferrari, C. and Canny, J. Planning optimal grasps. In *IEEE International Conference on Robotics and Automation*, pp. 2290–2295, 1992.

22. Goodwine, B. and Burdick, J.W. Motion planning for kinematic stratified systems with application to quasi-static legged locomotion and finger gaiting. *IEEE Trans. Robot. Autom.* **18**:209–222, 2002.

23. Gentilucci, M. and Corballis, M.C. From manual gesture to speech: A gradual transition. *Neurosci. Biobehav. Rev.* **30**:949–960, 2006.

24. Grioli, G., Catalano, M., Silvestro, E., Tono, S., and Bicchi, A. Adaptive synergies: An approach to the design of under-actuated robotic hands. *IEEE International Conference on Intelligent Robots and Systems (IROS)*, pp. 1251–1256, 2012.

25. Goldin, S. *Hearing Gesture: How Our Hands Help Us Think.* Cambridge, MA: Harvard University Press, 2003.

26. Haans, A. Mediated social touch: A review of current research and future directions. *Virtual Real.*, **9**(2):149–159, 2006.

27. Ritter, H., Haschke, R., Röthling, F., and Steil, J. Manual intelligence as a Rosetta Stone for robot cognition. *Robot. Res.* **66**:135–146, 2011.

28. Heidemann, G. and Schöpfer, M. Dynamic tactile sensing for object identification. In *IEEE International Conference on Robotics and Automation (ICRA)*, pp. 813–818, 2004.

29. Haschke, R., Steil, J., Steuwer, I., and Ritter, H. Task-oriented quality measures for dextrous grasping. In *IEEE CIRA Conference Proceedings*, pp. 689–694, 2005.

30. Han, L., Trinkle, J.C., and Li, Z.X. Grasp analysis as linear matrix inequality problems. *IEEE Trans. Robot. Autom.* **16**:663–674, 2000.

31. Maycock, J., Dornbusch, D., Elbrechter, C., Haschke, R., Schack, T., and Ritter, H. Approaching manual intelligence. *KI-Künstliche Intelligenz* **24**(4):287–294, 2010.

32. Jeannerod, M. The timing of natural prehension movements. *J. Mot. Behav.*, **16**:235–254, 1984.

33. Johansson, R.S. and Flanagan, J.R. Coding and use of tactile signals from the fingertips in object manipulation tasks. *Nature Rev. Neurosci.* **10**:345–359, 2009.

34. Hong, J., Lafferriere, G., Mishra, B., and Tan, X. Fine manipulation with multifinger hands. In *IEEE International Conference on Robotics and Automation*, pp. 1568–1573, 1990.

35. Steffen, J., Haschke, R., and Ritter, H. Towards dextrous manipulation using manipulation manifolds. In *IEEE/RSJ International Conference on Intelligent Robots and Systems (IROS)*, pp. 2738–2743, 2008.

36. Johansson, R.S. and Westling, G. Roles of glabrous skin receptors and sensorimotor memory in automatic control of precision grip when lifting rougher or more slippery objects. *Exp. Brain Res.* **56**:550–564, 1984.

37. Kawato, M. From understanding the brain by creating the brains towards manipulative neuroscience. *Phil. Trans. R. Soc. B*, 363:2201–2214, 2008.

38. Katz, D. and Brock, O. Extracting planar kinematic models using interactive perception. In D. Kragic and V. Kyrki (Eds.), *Unifying Perspectives in Computational and Robot Vision*, pp. 11–23. Springer US 2008.

39. Koiva, R., Haschke, R., and Ritter, H. Development of an intelligent object for grasp and manipulation research. In *Proceedings of Advanced Robotics (ICAR)*, pp. 204–210, 2011.

40. Klatzky, R. and Lederman, S. Stages of manual exploration in haptic object identification. *Percep. Psychophys.* **6**:661–670, 1992.

41. Liu, H. et al. Multisensory five-finger dexterous hand: The DLR/HIT hand. In *II. IEEE/RSJ International Conference on Intelligent Robots and Systems*, pp. 3692–3697, 2008.

42. Miller, A.T. and Allen, P.K. Graspit!: A versatile simulator for robotic grasping. *IEEE Robot. Autom. Mag.* **12**:110–122, 2004.

43. Morales, A., Asfour, T., Azad, P., Knoop, S., and Dillmann, R. Integrated grasp planning and visual object localization for a humanoid robot with five-fingered hands. In *IEEE/RSJ International Conference on Intelligent Robots and Systems (IROS)*, pp. 5663–5668, 2006.

44. Mittendorfer, P. and Cheng, G. Open-loop self-calibration of articulated robots with artificial skins. In *IEEE International Conference on Robotics and Automation (ICRA)*, pp. 4539–4545, 2012.

45. Ciocarlie, M., Lackner, C., and Allen, P. Soft finger model with adaptive contact geometry for grasping and manipulation tasks. In *IEEE 2nd Joint EuroHaptics Conference*, pp. 219–224, 2007.

46. Meier, M., Schöpfer, M., Haschke, R., and Ritter, H. A probabilistic approach to tactile shape reconstruction. *IEEE Trans. Robot.* **27**:630–635, 2011.

47. Salem, M., Kopp, S., Wachsmuth, I., Rohlfing, K., and Joublin, F. Generation and evaluation of communicative robot gesture. *Int. J. Social Robot.* **4**:1–17, 2012.

48. Lee, K.M., Jung, Y., Kim, J., and Kim, S.R. Are physically embodied social agents better than disembodied social agents?: The effects of physical embodiment, tactile interaction, and people's loneliness in humanrobot interaction. *Int. J. Hum. Comput. Stud.* **64** (10):962–973, 2006.

49. Namiki, A., Imai, Y., Ishikawa, M., and Kaneko, M. Development of a high-speed multi-fingered hand system and its application to catching. In *IEEE International Conference on Intelligent Robots and Systems (IROS)*, pp. 2666–2671, 2003.

50. Oztop, E., Kawato, M., and Arbib, M. Mirror neurons and imitation: A computationally guided review. *Neural Net.* **19**:254–271, 2006.

51. Prevete, R., Tessitore, G., Catanzariti, E., and Tamburrini, G. Perceiving affordances: A computational investigation of grasping affordances. *Cog. Syst. Res.* **12**:122–133, 2011.

52. Pulvermüller, F. Brain mechanisms linking language and action. *Nat. Rev. Neurosci.* **6**:576–582, 2005.

53. Rizzolatti, G. and Arbib, M. Language within our grasp. *Trends Neurosci.* **21**:188–194, 1998.

54. Rizzolatti, G. and Craighero, L. The mirror-neuron system. *Ann. Rev. Neurosci. (2004)*, **27**:169–192, 2004.

55. Detry, R., Ek, C.H., Madry, M., Piater, J., and Kragic, D. Generalizing grasps across partly similar objects. In *IEEE International Conference on Robotics and Automation (ICRA)*, pp. 3791–3797, 2012.

56. Röthling, F., Haschke, R., Steil, J., and Ritter, H. Platform portable anthropomorphic grasping with the bielefeld 20-dof shadow and 9-dof tum hand. In *IEEE/RSJ International Conference on Intelligent Robots and Systems (IROS)*, pp. 2951–2956, 2007.

57. Rosenbaum, D.A. and Jorgensen, M.J. Planning macroscopic aspects of manual control. *Human Move. Sci.* **11**:61–69, 1992.

58. Wells, R. and Greig, M. Characterizing human hand prehensile strength by force and moment wrench. *Ergonomics* **44**:1392–1402, 2001.

59. Scheibert, J. et al. The role of fingerprints in the coding of tactile information probed with a biomimetic sensor. *Science* **323**:1503–1506, 2009.

60. Schulz, S. First experiences with the Vincent hand. In *Proceedings of MyoElectric Controls/Powered Prosthetics Symposium*, Vol. 2011. New Brunswick (Canada), 2011.

61. Glover, S., Rosenbaum, D.A., Graham, J., and Dixon, P. Grasping the meaning of words. *Exp. Brain Res*, **154**:103–108, 2004.

62. Saxena, A., Driemeyer, J., and Ng, A.Y. Robotic grasping of novel objects using vision. *Int. J. Robot. Res.*, **27**:157–173, 2008.

63. Steffen, J., Elbrechter, C., Haschke, R., and Ritter, H. Bio-inspired motion strategies for a bimanual manipulation task. In *IEEE-RAS International Conference on Humanoid Robotics*, pp. 625–630, 2010.

64. Santello, M., Flanders, M., and Soechting, J.F. Postural hand synergies for tool use. *J. Neurosci.*, **18**:10105–10115, 1998.

65. Dahiya, R.S., Metta, G., Valle, M., and Sandini, G. Tactile sensing – from humans to humanoids. *IEEE Trans. Robot.* **26**(1):1–20, 2010.

66. Schöpfer, M., Pardowitz, M., Haschke, R., and Ritter, H. Identifying relevant tactile features for object identification. In E. Prassler (Ed.), *Towards Service Robots for Everyday Environments*, Vol. 76, pp. 417–430. Springer Berlin Heidelberg 2012.

67. Saut, J-P., Shabani, A., El-Khoury, S., and Perdereau, V. Dexterous manipulation planning using probabilistic roadmaps in continuous grasp subspaces. In *IEEE International Conference on Intelligent Robots and Systems (IROS)*, pp. 2907–2912, 2007.

68. Tunik, E. and Grafton, S.T. Beyond grasping: Representation of action in human anterior intraparietal sulcus. *TNeuroImage 36* **86**:77–86, 2007.

69. Schack, T. and Ritter, H. The cognitive nature of action – Functional links between cognitive psychology, movement science, and robotics. *Prog. Brain Res.* **174**:231–250, 2009.

70. Ückermann, A., Elbrechter, C., Haschke, R., and Ritter, H. 3d scene segmentation for autonomous robot grasping. In *IEEE/RSJ International Conference on Intelligent Robots and Systems (IROS)*, pp. 1734–1740, 2012.

71. Martinez-Hernandez, U., Lepora, N., Barron-Gonzalez, H., Dodd, T., and Prescott, T. Towards contour following exploration based on tactile sensing with the iCub fingertip. *Adv. Auton. Robot.* **7429**:459–460, 2012.

72. van Duinen, H. and Gandevia, S.S. Constraints for control of the human hand (2011). *J. Physiol.* **23**:5583–5593, 2011.

73. Chan, W.P., Parker C.A.C., Vander Loos H.F.M., et al. Grip forces and load forces in handovers: Implications for designing human-robot handover controllers. In *Proceedings of the 7th ACM/IEEE International Conference on Human-Robot Interaction*, pp. 9–16, 2012.

74. Yamazaki, Y., Yokochi, H., Tanaka, M., Okanoya, K., and Iriki, A. Potential role of monkey inferior parietal neurons coding action semantic equivalences as precursors of parts of speech. *Social Neurosci. 1*, **5**:105–117, 2010.

75. Yau, J.M., Pasupathy, A., Fitzgerald, P.J., Hsiao, S.S., and Connor, C.E. Analogous intermediate shape coding in vision and touch. *Proc. Nat. Acad. Sci*, **106**:16457–16462, 2009.

76. Su, Z., Fishel, J.A., Yamamoto, T., and Loeb, G.E. Use of tactile feedback to control exploratory movements to characterize object compliance. *Frontiers Neurorobot.* **6**, 2012.

4 Stochastic Information Processing that Unifies Recognition and Generation of Motion Patterns: Toward Symbolical Understanding of the Continuous World

Tetsunari Inamura and Yoshihiko Nakamura

CONTENTS

4.1 PREFACE

The bodies of humanoid robots recently show strides of advancement by adopting cutting-edge mechatronics and manufacturing technologies. This fact forms the background of the conducted research. It is necessary to found the technology to design a comparable information processing system. Namely, for building the brain of a humanoid, it has turned out that there are many essential problems that seem unsolvable based on the conventional analytic approaches of robotics. This research aimed to approach the essential problems of machine intelligence such as the whole-body motion pattern generation of large DOF systems, the dynamic interaction between the body and the world, embodied symbol emergence, and intelligence through emerged symbol manipulation, by carefully investigating and adopting the paradigms of neuroscience in a constructive way.

The research was titled "Development of Brain-Informatics Machines Through Dynamical Connection of Autonomous Motion Primitives" (PI: Yoshihiko Nakamura). It was conducted from December 1998 to November 2003, and financially supported by the Core Research for Evolutional Science and Technology (CREST) Program of the Japan Science and Technology Agency (JST) (Research Area: Creating the Brain / Research Supervisor: Shun-ichi Amari).

4.2 INTRODUCTION

Recently, human behavioral science and human intelligence have become conspicuous as a real robotics research issue. Although the motivation of artificial intelligence originated there, the physical limitations have forced or justified the researchers to carry on their research in a limited scope and scale of complexity. It ought to be the major challenge of contemporary robotics to study robotic behaviors and intelligence in the full scale of complexity, mutually sharing research outcomes and hypotheses with human behavioral science and human intelligence.

The discovery of mirror neurons [1] has been a notable topic of brain science. These neurons have been found in primate and human brains, and fire when the subject observes a specific behavior and also fire when the subject starts to act on the same behavior. Furthermore, it is located on Broaa's area which has a close relationship with language management. The fact suggests that the behavior recognition process and behavior generation process are combined as the same information processing scheme, and the scheme is nothing but a core engine of symbol manipulation ability.

Indeed, in Donald's "Mimesis theory" [2] [3], it is said that symbol manipulation and communication ability are founded on behavior imitation, that is, integration of behavior recognition and generation. We believe that a paradigm can be proposed taking advantage of the mirror neurons, with considerations of Deacon's contention [4] that language and brain had evolved simultaneously.

For this purpose, we have developed the following four frameworks: (1) a mutual connection model between motion patterns and symbols based on a hidden Markov model, (2) keyframe compression and decompression for time-series data based on the continuous hidden Markov model, (3) imitation learning model with embodiment based on discrete continuous hybrid HMM, and (4) development and manipulation of proto-symbols based on geometric proto-symbol space.

4.3 MUTUAL CONNECTION MODEL BETWEEN MOTION PATTERNS AND SYMBOLS BASED ON HIDDEN MARKOV MODELS

We have focused on a stochastic information processing framework of the hidden Markov model (HMM) in order to integrate symbol representation and motion patterns of humanoid robots that have many degrees-of-freedom. The HMM is regarded as a symbol representation called a *proto-symbol*, and is also used for the development of a mutual connection model. The mutual connection model consists of two phases. In the first half, observed motions are transformed into motion elements by comparing, and are abstracted as proto-symbols. Observed motion patterns are analyzed into the motion elements, and the sequences of the motion elements are abstracted into proto-symbols, regarded as a behavior series. Figure 4.1 shows the HMM which expresses motion patterns, that is, time-series data of motion elements.

Generation of motion patterns from the HMM is equal to the generation of time-series data of motion elements. However, it is difficult to calculate the sequence using

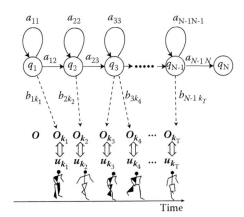

FIGURE 4.1   Discrete hidden Markov models and humanoids' motion.

only the HMM, because the generation process also searches a motion pattern that has the best likelihood value among all the motion patterns. The simplest way to generate suitable motion patterns is to find the maximum likelihood by scanning the entire pattern space. However, it is difficult to adopt this method because the size of the search space will increase in proportion to the exponential of the time length of the motion pattern. In order to encode a motion pattern into a chromosome, each motion element corresponds to each gene. As the fitness of the chromosome, the likelihood of that the motion patterns are generated by the HMM is used. It has also adopted translocation, not simply crossover and mutation. It is suitable for the evolution to keep a behavior series because the block of self-motion elements indicates the behavior series.

A mathematical model for the integration of motion recognition and generation is achieved using the above methods. Figure 4.2 shows an outline of the proposed framework.

4.4 KEYFRAME COMPRESSION AND DECOMPRESSION FOR TIME-SERIES DATA BASED ON CONTINUOUS HIDDEN MARKOV MODELS

Memory of motion patterns such as data, comparison of a new motion pattern with the data, and playback of one from the data are inevitably involved in the information processing of intelligent robot systems. Such computation forms the computational foundation of learning, acquisition, recognition, and generation processes of intelligent robotic systems. Motion patterns along with temporal sensory data would be appropriate to describe behaviors of a robot. This is the computational problem of time-series data and the subject of the present chapter.

The computational problem of time-series data would need to consider: (1) efficiency of data compression/decompression, and (2) unification of algorithms for memory (compression), comparison, and playback (decompression). The former is mandatory because it determines the volume of the database of motion patterns. The latter is not a must, but an important requirement to maintain consistency of the three kinds of computation. We have focused on the keyframe representation for this purpose. *Keyframe* is one of the famous methods for motion design, especially in computer graphics, which is a motion pattern at several impressive moments. Time-series data of motion patterns are combined using these computer graphics keyframes. Recently, the keyframe has been used for motion planning of robots and motion recognition because the frame representation has an affinity for such issues.

The proposed mimesis model has no criterion for designing the motion element. It is effective for the mimesis model to adopt the keyframe representation for the motion elements design.

For the keyframe representation, a continuous hidden Markov model (CHMM) is used as shown in Figure 4.3. The CHMM consists of a finite set of states $S = \{q_1, \ldots, q_N\}$, a state transition probability matrix $A = \{a_{ij}\}$ (the probability of state transition from q_i to q_j), an output probability matrix $B = \{b_{ij}\}$, and an initial distribution vector $\pi = \{\pi_i\}$, that is, a set of parameter $\lambda = \{Q, A, B, \pi\}$.

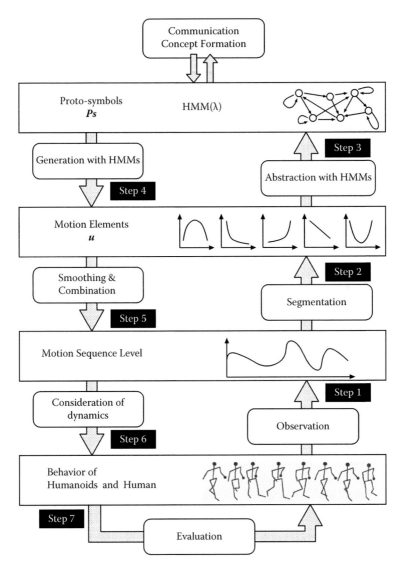

FIGURE 4.2 Mutual connection model between motion patterns and symbols based on hidden Markov models.

Here, $\{b_{ij}\}$ is the output probability density function

$$b_i(o) = \sum_{j=1}^{M} c_{ij} \mathcal{N}_{ij}(o; \mu_{ij}, \Sigma_{ij}) \qquad (4.1)$$

that relates continuous output vector o with the ith state node q_i. M indicates the number of Gaussian functions as $\mathcal{N}(o; \mu, \Sigma)$. The CHMM generates motion patterns by the stochastic process shown in Figure 4.3. We have defined the motion elements

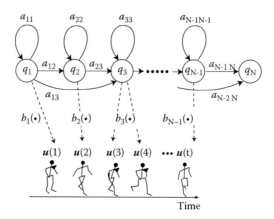

FIGURE 4.3 Continuous hidden Markov models and humanoid's motion.

as keyframes of the motion pattern as follows,

$$u \stackrel{\text{def}}{=} \{\mu, \Sigma\} \tag{4.2}$$

These parameters are calculated by the Baum–Welch algorithm.

Here, there are fluctuations among each reproduction process because of the property of stochastic models. The time length of the motions always changes by the state transition probability A; the value of each moment of the time-series data also always changes by the output probability $b_i(o)$. We thus propose an average strategy in order to cancel these fluctuations:

*step*1 Getting a state transition sequence
$$Q = \{q_{k[1]}, q_{k[2]}, \ldots, q_{k[T]}\},$$
$$(k[i] \in \{1, 2, \ldots, N\})$$
with a trial of the state transition process;

*step*2 An average of transition sequence $\hat{Q}$ is
calculated by n_q times repetition
$(Q_1, \ldots, Q_{n_q})$ of *step*1;

*step*3 Output time-series data O is calculated by
a trial of output according to the average transition
sequence $\hat{Q}$;

*step*4 $O_1, \ldots, O_{n_o}$ is calculated by
n_o times repetition from *step*1 to *step*3;

*step*5 Average time-series data $\hat{O}$ is
calculated by the $O_1, \ldots, O_{n_o}$ after regularization of the time length;

where n_q, n_o are decided experimentally.

Figure 4.4 shows the example decomposition result using the method. The target data are a certain joint angle of a humanoid robot. The dotted line indicates the original time-series data; the dashed line indicates a result of a single output trial (Q_i). There are deviances for time direction and value direction, however; the deviance for time

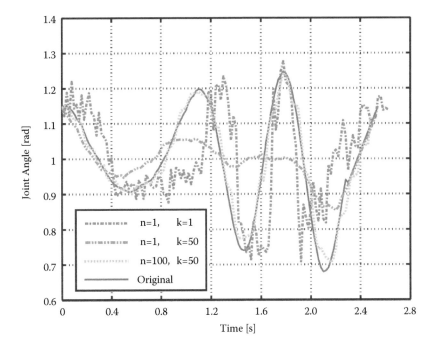

FIGURE 4.4   Generation of time-series data using continuous HMM.

direction was cancelled using *step2*, and the deviance for value direction was also cancelled using *step5*. The final output result is shown as a solid line. As the figure shows, output time-series data are similar to the original time-series data.

4.5   IMITATION LEARNING MODEL WITH EMBODIMENT BASED ON DISCRETE/CONTINUOUS HYBRID HMM

There were several remaining problems in the mimesis model as follows:

1. There is no exact principle of how to design the motion elements. In previous works, static and limited motion elements had been embedded by the developer beforehand. It is necessary for the mimesis system to develop the motion elements in order to be suitable for imitation learning from observational experience.

2. Physical condition of the motion had not been taken into consideration. As physical characteristics of the learner and demonstrator are different, the observer therefore cannot reproduce the same motion. The motion elements have to be suitable for both recognition of other fs motion and embodiment of humanoids.

3. The motion elements corresponded to each joint. Thus the number of motion elements became no less than the number of DOFs in order to represent the whole body motion. It caused the complexity of symbol representation.

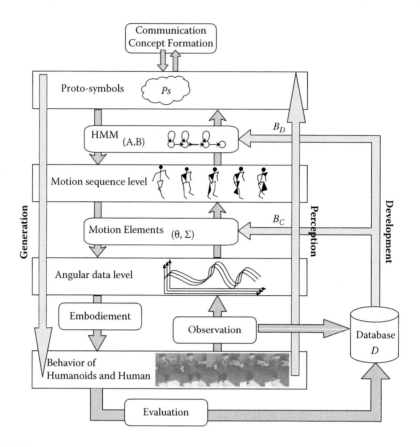

FIGURE 4.5   Hybrid mimesis model.

Furthermore, the motion elements should represent correlation information between each joint.

For the above-mentioned reasons, we propose a new method for acquisition of motion elements with two characteristics: (a) use of continuous HMM and (b) modification of elements during observation and the generation loop. Approach (a) enables the system to avoid problems (1) and (2). On the other hand, approach (b) enables the system to avoid problems (2) and (3).

For our purposes, continuous HMM (CHMM) was introduced which can treat continuous data. Although many advantages are available, CHMMs have a disadvantage in that huge computational quantity is needed and would take a long time for motion generation and recognition. Therefore we have proposed a hybrid hidden Markov model that consists of CHMM and discrete HMM (DHMM) as shown in Figure 4.5. In the motion recognition and generation phases, DHMM is used, whose computational quantity is low. In the motion elements acquisition phase, CHMM is used, whose computational quantity is high.

The mimesis model acquires adequate motion elements during the repetition of motion recognition and generation. The following steps constitute the development

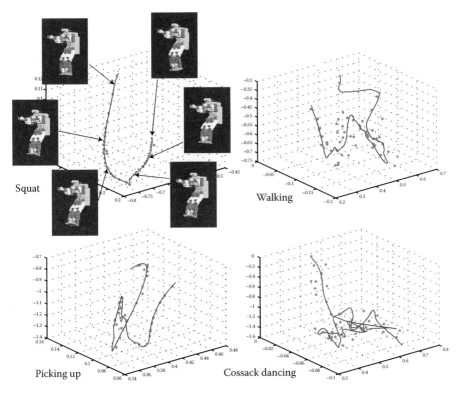

FIGURE 4.6   Acquired key frame for four motions.

procedure of motion elements:

1. Generation of the motion patterns from proto-symbols.
2. Evaluation of the motion patterns based on recognition criterion.
3. If the score is good, the motion patterns are added into the database.
4. Return to **step1**, after the recalculation of motion elements.

The performance of motion recognition and generation is influenced by the characteristic of motion elements. If the motion elements had no relationship between the observed motion, the recognition process would fail. Therefore, we adopted an approach that the system searches the best motion elements with an evaluation criterion of whether the generated motion would be fit for the body and the recognition would succeed against familiar motions. Using the method, the humanoid can acquire adequate motion elements through repetition of motion perception and generation.

After 50 observations, the motion generation process is executed. A result with the limitation condition is shown in Figure 4.6. In the figure, three axes indicate hip joint (pitch), and knee joint and ankle joint (pitch). The curved line in the figure corresponds to the motion trajectory. The dots indicate acquired motion elements. The result shows that the motion elements are basically located near the original motion trajectory.

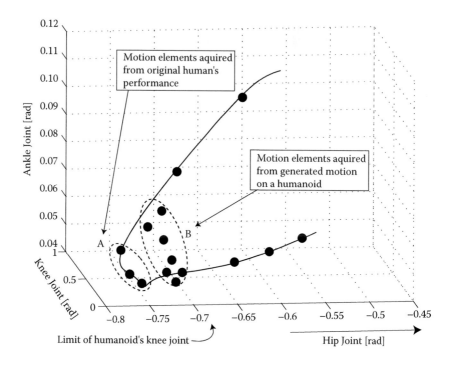

FIGURE 4.7   Acquired embodied self-motion elements using loop structure.

Additionally, the motion elements gathered not only around the A area, but also around the B area in Figure 4.7. The motion elements located around the B area are acquired by the generated self-motions in the database, which fit for the humanoid embodiment. This result shows that both motion elements are acquired: elements for the recognition of others' motions (A area) and ones for the generation of self-motion (B area).

4.6 DEVELOPMENT AND MANIPULATION OF PROTO-SYMBOLS BASED ON GEOMETRIC PROTO-SYMBOL SPACE

4.6.1 HIERARCHICAL MIMESIS MODELS

Here, one defect arises that the relationship between two similar behaviors has been lost when the behavior is transferred into proto-symbols. An aspect of the symbol is that a semantic relation exists among symbols, contrary to the icons and labels that have no relation representation between each element. To solve this defect, a hierarchical structure is needed in which the relation in the lower motion pattern layer will be kept in the higher symbol layer. We call such a structure the *hierarchical mimesis model*.

To construct such a hierarchical model, the mathematical framework should have an ability to treat the distance relationship between each behavior and symbol, and to

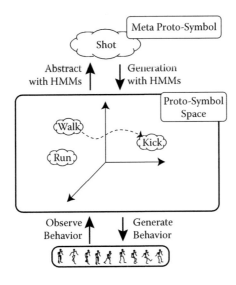

FIGURE 4.8   Outline of hierarchical mimesis model.

represent the hierarchical structure. We call such a structure *proto-symbol space*. A conceptual image of such a hierarchical mimesis model is shown in Figure 4.8. In the model, each proto-symbol is represented by a continuous vector; that is, continuous proto-symbol space exists above the low-level behavior pattern. Using the proto-symbol space, the human's long-term behavior is transferred into a trajectory in the space. We can also abstract the trajectory as a symbol–like representation using the same HMM. In other words, a hierarchical abstract framework can be established easily based on the HMM. Continuous HMM is the adequate way to realize such a hierarchical structure.

4.6.2 DEFINITION OF DISTANCE BETWEEN HMMs

Although distance is needed for construction of space, distance between two HMMs is not able to be defined easily because it is a stochastic model. For such stochastic models, there is a method to express the distance information. In this chapter, we adopt Kullback–Leibler information as the representation of distance between HMMs. To be more precise, the Kullback–Leibler information is not distance because it does not satisfy the property of the distance, triangle inequality and symmetry, therefore we call the Kullback–Leibler information a degree of similarity of the HMMs. The Kullback–Leibler information against two stochastic models p_1 and p_2 is defined as

$$D(p_1, p_2) = \int_{-\infty}^{\infty} \left(p_1(x) \log \frac{p_1(x)}{p_2(x)} \right) dx \qquad (4.3)$$

To apply Equation (4.3) to HMMs, the following equation is usually used [5].

$$D(\lambda_1, \lambda_2) = \sum_n \frac{1}{T_n} \left[\log P(y_1^T | \lambda_1) - \log P(y_1^T | \lambda_2) \right] \qquad (4.4)$$

As Equation (4.4) does not satisfy the distance axiom, we use following improved information,

$$Ds(\lambda_1, \lambda_2) = \frac{1}{2} (D(\lambda_1, \lambda_2) + D(\lambda_2, \lambda_1)) \qquad (4.5)$$

4.6.3 CONSTRUCTION OF SPACE BASED ON SIMILARITY

In order to construct proto-symbol space from the distance information, multidimensional scaling (MDS) is used. MDS is a method that accepts distance information among elements and outputs the position of each element in the generated space. Let the similarity between the ith element and the jth element be f_{ij}, and the distance between the ith and the jth elements be d_{ij}. MDS makes the following error the minimum for space construction:

$$S^2 = \sum_{i,j} (f_{ij} - d_{ij})^2 \qquad (4.6)$$

In the case of HMMs, Kullback–Leibler information Ds is used for the similarity f_{ij}. Let the position of each HMM be $x = \{x_1, x_2, \ldots, x_n\}$, where n is the number of dimensions of the space. Using the least-squares method, each position x is calculated.

We confirmed the performance of the proposed space construction method against six motion patterns as shown in Figure 4.11. At first, we gave a ten-dimensional vector for each $\{x_1, x_2, \ldots, x_n\}$. Figure 4.9 shows the constructed space and location of each proto-symbol (HMM). As the diagram indicates, the first to the fourth dimensions are effectively used for the space construction; however, the rest of the dimensions are not well used. Therefore, we adopted a three-dimensional proto-symbol space as shown in Figure 4.10.

After the construction of the proto-symbol space, a certain state point in the proto-symbol space indicates a motion pattern. It means that the state points can reproduce the original motion patterns. Figure 4.12 shows the motion reproduction result through the proto-symbol space shown in Figure 4.10. Even though the 3D proto-symbol space skipped several dimensions, the generated motion patterns are similar to the original human motion. It supports the adequacy of the construction method of the proto-symbol space.

4.7   BEHAVIOR MANIPULATION BY PROTO-SYMBOL MANIPULATION

4.7.1 PROTO-SYMBOL MANIPULATION

Symbol manipulation for the motion pattern of humans and humanoid robots is defined as follows:

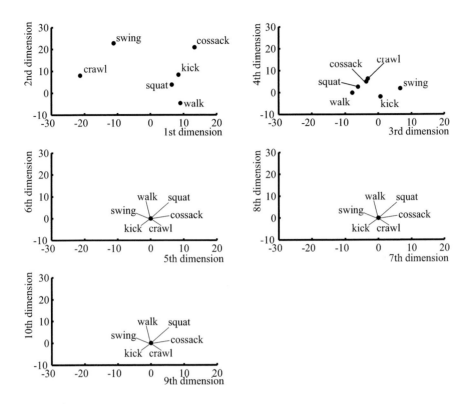

FIGURE 4.9 Result of proto-symbol space.

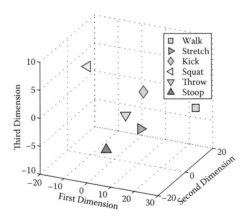

FIGURE 4.10 Six motions in 3D proto-symbol space.

(a)

(b)

(c)

(d)

(e)

(f)

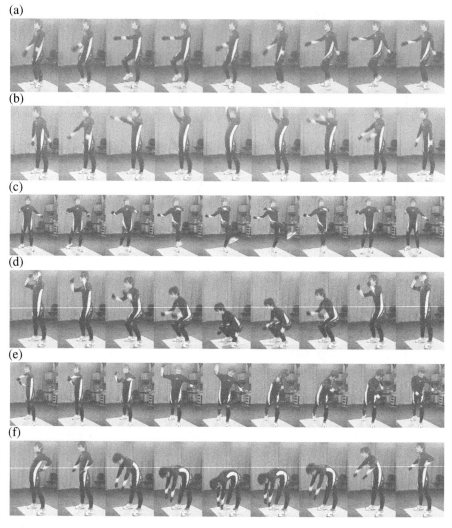

FIGURE 4.11   Six motions performed by humans.

- Generation of novel behavior using known basic motions
- Recognition of novel behavior using known basic motions
- Abstract of novel behavior using known basic motions

From the viewpoint of the proto-symbol space, the above definitions can be interpreted as the following:

- Generation: Creation of a novel state point and motions using existing state points of proto-symbols
- Recognition: Recognization of a novel sequence of state points using the existing state points of proto-symbols

(a)

(b)

(c)

(d)

(e)

(f)

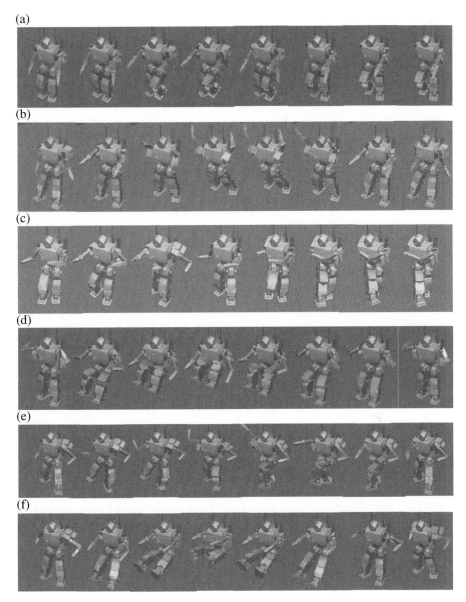

FIGURE 4.12 Six motions imitated by humanoid.

- Abstract: Abstract of a novel sequence of state points into meta-proto–symbol

A mutual conversion method between motion patterns and symbol representation in proto-symbol space has been established in the previous section. In this section, we discuss a proto-symbol manipulation method where generation and recognition of novel motion are defined by a simple structure.

4.7.2 GENERATION OF NOVEL PROTO-SYMBOL

Creating a novel proto-symbol is equal to creating a novel state point on the proto-symbol space. To create a novel state point, the following composition regulation is used,

$$b_i(o) = \sum_{m=1}^{M} \alpha c_{im_A} \mathcal{N}(\mu_{im_A}, \sigma_{im_A})$$

$$+ \sum_{m=1}^{M} (1-\alpha) c_{im_B} \mathcal{N}(\mu_{im_B}, \sigma_{im_B}) \qquad (4.7)$$

$$a_{ij} = \alpha a_{ij_A} + (1-\alpha) a_{ij_B} \qquad (4.8)$$

Equations (4.7) and (4.8) are applied when a novel state point is located on a straight line connecting the two points (λ_A and λ_B). When a novel state point is not on any straight lines connecting known proto-symbols, the parameter is composed according to the distance ratios among each known proto-symbol as follows,

$$b_i(y) = \sum_{m=1}^{M} \frac{1}{d_l \sum_l \frac{1}{d_l}} c \mathcal{N}(\mu_{im}^l, \rho_{im}^l)$$

$$a_{ij} = \sum_{m=1}^{M} \frac{1}{d_l \sum_l \frac{1}{d_l}} a_{ij}^l \qquad (4.9)$$

where d_l is the distance between a novel state point and known proto-symbol λ_l. Finally, a low-level motion pattern is generated from the state point, using the method proposed in Inamura, Tanie, and Nakamura [6].

4.7.2.1 Novel Motion Generation from State Sequence in Proto-Symbol Space

Furthermore, motion generation can be performed against the state transition sequence in the proto-symbol space. In this section, a motion generation method in which a state transition sequence is given.

Let the state transition be $x[1], x[2], \ldots, x[n]$. A generation method in which a fixed state point is given in the proto-symbol space was introduced before; the continuous generation by transitional state points is equivalent to the average of motions generated by those state points.

Figure 4.14 shows the outline of the generation process. In step 1, motion patterns are generated from each state point in the proto-symbol space using the proposed method [6]. In step 2, the time length of all motion patterns is set to the same value T_c in order of composition. In steps 3 and 4, partial motion patterns are picked up based on the phase information for each state point, that is, the charging period of time for each state point. Finally in step 5, composite motion patterns are generated.

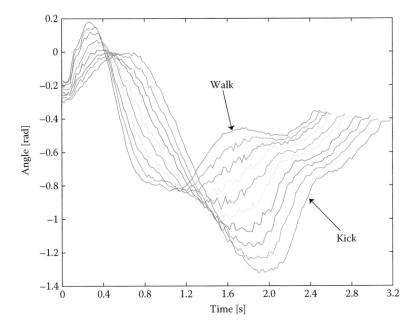

FIGURE 4.13 Generated motion from mixed HMMs.

4.7.3 Recognition of Novel Motion Based on Proto-Symbol Manipulation

The proto-symbol–based recognition is equivalent to deciding a state point in the proto-symbol space. After the decision, the relationship between proto-symbols brings about the symbol representation of the novel motion.

The difference between the observed motion O and each proto-symbol is represented as the likelihood $P(O|\lambda_i)$. To convert the likelihood into distance information, the following equation

$$D(\lambda_k, \lambda_i) = \sum \frac{1}{T} \left(\log P(o_k^T |\lambda_k) - \log P(o_k^T |\lambda_i) \right) \qquad (4.10)$$

is used; however, the $P(O|\lambda_O)$ is not decided before learning of the HMM λ_O. Here, we introduce an approximation for the $P(O|\lambda_O)$.

Table 4.1 shows the logarithm of the likelihood for each motion and proto-symbol. Diagonal values indicate from about 5 to 20; however, the rest show very little value. According to the result, we had an assumption that a novel behavior also follows the above empirical rule that

$$\log P(O|\lambda_O) = 20 \qquad (4.11)$$

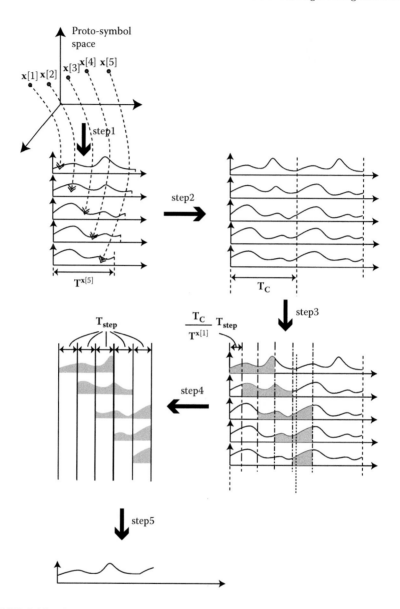

FIGURE 4.14   Outline of motion generation.

4.7.3.1   Novel Motion Recognition as State Sequence in Proto-Symbol Space

The outline of novel motion recognition as a state sequence in the proto-symbol space is shown in Figure 4.15.

In step 1, focusing on the period of time T_{span} in the observed motion pattern $\boldsymbol{O} = [\, \boldsymbol{o}_1 \quad \boldsymbol{o}_2 \quad \cdots \quad \boldsymbol{o}_T \,]$. Let the cut-off motion pattern be $\boldsymbol{O}_1 = [\, \boldsymbol{o}[1] \quad \boldsymbol{o}[2] \quad \cdots \quad \boldsymbol{o}[T_{span}]\,]$. In step 2, the state point is decided using the method in the previous paragraph. Next, shift the focus point, and let the kth focus point be $\boldsymbol{O}_k = \{\, \boldsymbol{o}_{1+(k-1)\cdot T_{step}}, \dots,$

TABLE 4.1
Likelihood

	Dance_hmm	Crawl_hmm	Kick_hmm	Squat_hmm	Swing_hmm	Walk_hmm
Dance	15.00	−11609	−3978	−4501	−5471	−6736
Crawl	−12833	19.03	−9329	−7864	−5985	−10126
Kick	−4526	−9605	13.41	−2901	4985	4312
Squat	−6624	−9248	−3239	7.05	−8278	−2021
Swing	−8810	−6433	−7762	−9339	18.86	−13469
Walk	−8573	−10776	−4498	−1021	−10944	5.49

$o_{1+T_{span}+(k-1)\cdot T_{step}}\}$, with an increase of the index as $k = 1, 2, \ldots, ((T - 1 - T_{span})/T_{step}) + 1$. Finally in step 3, the sequence of state points in the proto-symbol space is acquired.

4.8 EXPERIMENTS

4.8.1 RECOGNITION OF NOVEL MOTION

First, recognition in which motion patterns are transferred into a state point in proto-symbol space was performed. Two novel motions "throwing with kicking" and "stretching with walking" are target motions. Figure 4.16 shows the recognition result. The squares and triangles are known basic proto-symbols. The small dots indicate result state points. As the diagram shows, two dot marks are located on the line between each basic proto-symbol.

Next, recognition in which novel motion patterns are transferred into a sequence of state points in the proto-symbol space was performed. The target motion is "walking first, then shift to kicking." The result is shown in Figure 4.17. As the diagram shows,

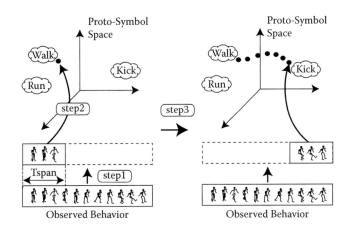

FIGURE 4.15 Procedure of projecting motion in proto-symbol space.

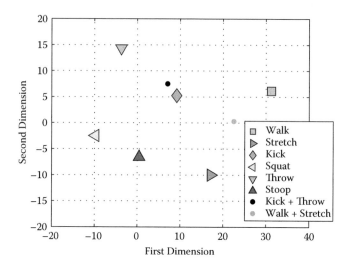

FIGURE 4.16   A recognition result of novel motions.

recognized dot marks start from the proto-symbol of "walk", and end at the proto-symbol of "kick."

4.8.2   GENERATION OF NOVEL MOTION

In this experiment, we have investigated the motion output when a trajectory is given in the proto-symbol space. As a given trajectory, we prepared a simple line trajectory from the "walking" state point to the "kicking" state point. Figure 4.18 is the result

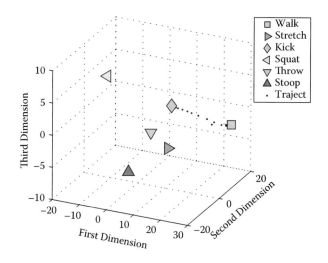

FIGURE 4.17   A result of motion recognition in the proto-symbol space.

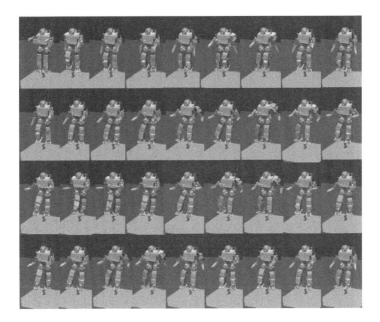

FIGURE 4.18   Generated motion: from walking to kick.

of motion output. As the figure shows, the humanoid motion is adequately controlled as the symbol manipulation in the proto-symbol space.

4.9   DISCUSSION

4.9.1   COMPARISON WITH CONVENTIONAL RESEARCH

Motion recognition using the HMM is a famous method, thus much research has been proposed for gesture recognition or behavior understanding [7–10]; however, no research has existed in which motion is generated from HMMs. Masuko et al. [11] [12] have proposed a speech parameter generation method using HMMs, but the generation process is not the opposite of the speech recognition process. The most important characteristic of our method is that the motion recognition and motion generation processes are integrated by only a HMM.

Acquisition and symbolization of the motion patterns of robots have been discussed in Doya and coworkers' research [13] [14]. In their MOSAIC model, time-series data are acquired as a sequence of symbol representations; however, the symbol representations do not contain dynamics information in the time-series data. In contrast, our method, the proto-symbol representation, contains dynamics information in motion patterns. Another issue different from Doya's research is that the single HMM acts as a recognition and generation model. In Doya's MOSAIC model, two modules, a prediction module (for generation) and a control module (for recognition), become companions as a motion primitive. We think that single HMM is a better choice for abstract symbol representation.

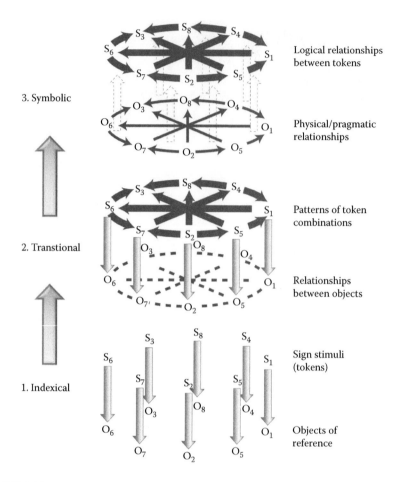

FIGURE 4.19   Development of symbol representation from indexical level by Deacon. S indicates symbol; o indicates raw-level pattern such as a motion pattern.

To integrate the recognition and generation processes for time-series data, some dynamical approaches have been proposed [15] [16]. In this research, an explicit dynamics–like attractor is designed in order to abstract the time-series data. In conventional research, a recurrent neural network is one of the effective ways for the dynamics design [17–20]. However, it is difficult to design symbol representation if these methods are adopted as the dynamics representation. The HMM is effective for both dynamics representation and symbol representation.

Symbol emergence has been tried in conventional artificial intelligence research. The most difficult issue of symbol emergence is how to manipulate the created symbol representation, contrary to the ease of symbol creation. Deacon has proposed the symbol development model [4] as shown in Figure 4.19. In his theory, symbolic representation is developed from the indexical level and iconic level. In the indexical level, a simple relationship between a motion pattern and a symbol representation is established; however, the relationship between each symbol and motion pattern is not considered. In the transitional level, the relationship between each symbol is developed as a token combination, then the relationship between each motion pattern starts to be constructed. In the final level, the logical relationship between the symbols is combined with the physical relationship between motion patterns. Our approach follows the development model. At the present moment, our method achieved the transitional level and is going to achieve the final symbolic level.

4.10 CONCLUSION

To solve the defect of the usual mimesis model, that is, the impossibility of symbol manipulation, we introduce a hierarchical mimesis model. The hierarchical model is built on a continuous hidden Markov model and distance representation using a multidimensional scaling method. The continuous hidden Markov model enables the model to generate the humanoid's motion naturally as the contrary direction of motion recognition. The multidimensional scaling method enables the model to describe the relationship between each proto-symbol, namely continuous HMM. Owing to the distance representation, symbol manipulation is achieved as geometric state manipulation. Through experiments, the following abilities are realized: (1) novel motion can be recognized as a combination of the known motion's proto-symbols, and (2) novel motion can be generated by a combination of known proto-symbols.

The remaining problem is the defectiveness of proto-symbol space. Even if the median point in the proto-symbol space between two proto-symbols generates natural motions, the characteristics of the motion are not always similar to the characteristics of two real motions. The most frequent cause is that proto-symbol space is not Euclidean space. Toward this issue, we believe that information geometry would achieve the desired effect.

ACKNOWLEDGMENT

This research was supported by the Core Research for Evolutional Science and Technology (CREST) program of the Japan Science and Technology Corporation (PI: Y. Nakamura).

REFERENCES

1. Gallese, V. and Goldman, A. Mirror neurons and the simulation theory of mind-reading. *Trends Cogn. Sci.* **2**, (12):493–501, 1998.
2. Donald, M. *Origins of the Modern Mind.* Cambridge, MA: Harvard University Press, 1991.
3. Donald, M. Mimesis and the Executive Suite: Missing Links in Language Evolution. In J. Hurford, M. Kennedy, and C. Knight (Eds.), *Approaches to the Evolution of Language: Social and Cognitive Bases*, Cambridge, UK: Cambridge University Press, Ch. 4, pp. 44–67, 1998.
4. Deacon, T.W. *The Symbolic Species.* New York: W.W. Norton, 1997.

5. Rabiner, L.R. and Juang, B.H. A probabilistic distance measure for hidden Markov models. *AT&T Tech. J.* **1**, 64:391–408, 1985.

6. Inamura, T., Tanie, H., and Nakamura, Y. Keyframe compression and decompression for time series data based on the continuous hidden Markov model. In *Proceedings of International Conference on Intelligent Robots and Systems*, pp. 1487–1492, 2003.

7. Ogawara, K., Takamatsu, J., Kimura, H., and Ikeuchi, K. Modeling manipulation interactions by hidden Markov models. In *Proceedings of 2002 IEEE/RSJ International Conference on Intelligent Robots and Systems*, pp. 1096–1101, 2002.

8. Pook, P.K. and Ballard, D.H. Recognizing teleoperated manipulations. In *IEEE International Conference on Robotics and Automation*, pp. 578–585, 1993.

9. Wada, T. and Matsuyama, T. Appearance based behavior recognition by event driven selective attention. In *IEEE Computer Society Conference on Computer Vision and Pattern Recognition*, pp. 759–764, 1998.

10. Yamato, J., Ohya, J., and Ishii, K. Recognizing human action in time-sequential images using hidden markov model. In *IEEE Computer Society Conference on Computer Vision and Pattern Recognition*, pp. 379–385, 1992.

11. Kobayashi, T., Masuko, T., Tokuda, K., and Imai, S. Speech synthesis from hmms using dynamic features. In *Proceedings of International Conference on Acoustics, Speech, and Signal Processing*, pp. 389–392, 1996.

12. Imai, S., Tokuda, K., and Kobayashi, T. Speech parameter generation from hmm using dynamic features. In *Proceedings of International Conference on Acoustics, Speech, and Signal Processing*, pp. 660–663, 1995.

13. Doya, K., Samejima, K., Katagiri, K.I., and Kawato, M. Multiple model-based reinforcement learning. *Neural Comput.* **14**, 1347–1369, 2002.

14. Samejima, K., Doya, K., and Kawato, M. Inter-module credit assignment in multiple model-based reinforcement learning. *Neural Netw.* **16**, (7):985–994, 2003.

15. Okada, M., Tatani, K., and Nakamura, Y. Polynomial design of the nonlinear dynamics for the brain-like information processing of whole body motion. In *Proceedings of IEEE International Conference on Robotics and Automation*, pp. 1410–1415, 2002.

16. Ijspeert, A.J., Nakanishi, J., and Schaal, S. Movement imitation with nonlinear dynamical systems in humanoid robots. In *Proceedings of IEEE International Conference on Robotics and Automation*, pp. 1398–1403, 2002.

17. Tani, J. On the dynamics of robot exploration learning. *Cogn. Syst. Res.* **3**, (3):459–470, 2002.

18. Inamura, T., Nakamura, Y., and Simozaki, M. Associative computational model of mirror neurons that connects missing link between behaviors and symbols. In *Proceedings of International Conference on Intelligent Robots and Systems*, pp. 1032–1037, 2002.

19. Morita, M. Memory and learning of sequential patterns by nonmonotone neural networks. *Neural Netw.* **9**, (8):1477–1489, 1996.

20. Morita, M. and Murakami, S. Recognition of spatiotemporal patterns by nonmonotone neural networks. In *Proceedings of the 1997 International Conference on Neural Information Processing*, Vol. 1, pp. 6–9, 1997.

5 Foveal Vision for Humanoid Robots

Ales Ude

CONTENTS

Human vision is based on the ability to perceive light, which enters the brain through the eye as shown in Figure 5.1. Light travels through a number of layers before hitting the area of the eye called the retina [16]. It is converted into nerve signals in photosensitive cells, which primarily consist of cones and rods. Cones are color-sensitive and require a lot of light to be triggered. They are most densely distributed in the foveal area of the retina, which is responsible for detailed vision. There are no rods in the the fovea. Rods are more densely distributed in peripheral areas of the retina, which contain significantly fewer cones. They are much more sensitive to light than cones and are therefore responsible for vision under low-light conditions. They are monochromatic and are also responsible for motion detection. Image resolution in the periphery is much lower than in the fovea also due to the pooling of information from retinal receptors by retinal ganglion cells, which is far greater in the visual periphery than in the foveal area [18]. The distribution of cones and rods on the retina takes into account the competing evolutionary requirements for a wide field of view and high-resolution vision. The light is carried into the brain through the ganglion cells and the optic nerve, which enters the retina at the optical disc.

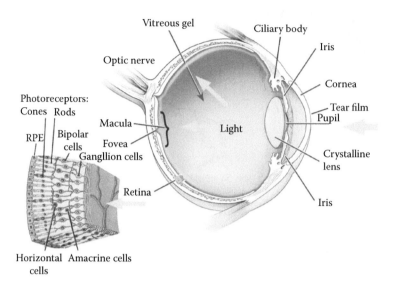

FIGURE 5.1   The anatomy of human eye and the perception of light. (Adopted from Vera-Diaz and Doble, 2011, where published under license http://creativecommons.org/licenses/by/3.0/. With permission.)

Each eye has six extraocular muscles attached to it and moves because the appropriate ones shorten [16]. There exist several types of eye movements, but in the context of foveated vision, the most interesting ones are *saccades* and *smooth pursuit*. A saccade is a rapid, goal-directed eye movement to bring the area of interest to the area of highest resolution, (i.e., the fovea [11]). Smooth pursuit eye movements are slower. Their task is to keep the object of interest in the fovea and stabilize its image [15,31]. Combined, they thus enable the processing of high-resolution images of the observed object. Both saccades and smooth pursuit eye movements can be complemented by the head to expand their range [31].

Object and face recognition rely on detailed analysis of foveal images [19]. Thus to improve the capabilities of artificial vision systems, which are still far less capable than human vision, it is necessary to develop active object recognition systems that enable the processing of high-resolution foveal images. On the other hand, it has been shown that for tasks such as recognition of the scene gist [18] and place recognition [19], peripheral vision is all that is needed. Active systems that enable simultaneous acquisition of variable resolution images should thus be developed to improve the performance of artificial vision systems.

Various computational models have been proposed to explain the high performance of human object recognition. Most notable among them are structural description models [3] and viewpoint-dependent models [30], but a comprehensive theory of object recognition is still elusive. fMRI has significantly changed the research on visual recognition and resulted in new findings, for example, with respect to category selectivity [25]. However, new neuroscientific results have not yet resulted in well-specified computational models. The fields of computer vision, where statistical approaches are by far most successful to date, and theories of human object perception are still developed largely independently of each other.

This chapter is concerned with the development of a foveated visual system for object recognition on a humanoid robot. First we discuss active humanoid vision systems that realize foveated vision. An analysis showing the theoretical degree of precision that can be achieved by a setup with two cameras in each eye is presented. We describe a control scheme that can be used to implement smooth pursuit movements and maintain the view of the observed object in the foveal images using information from the peripheral views. We show that the proposed foveation control system combined with a suitable tracker can be utilized for object learning and recognition. Our experimental results demonstrate that higher-resolution images provided by foveated vision result in better classification rates.

5.1 ARTIFICIAL FOVEATED VISION SYSTEMS

System designs proposed to mimic the foveated structure of biological vision include systems with two cameras in each eye [1,2,5,8,17,28] (i.e., a narrow-angle foveal camera and a wide-angle camera for peripheral vision); lenses with space-variant resolution [26] (i.e., a very high-definition area in the fovea and a coarse resolution in the periphery); and space-variant log-polar sensors with retinalike distribution of photo-receptors [27]. It is also possible to implement log-polar sensors by transforming standard images into log-polar ones [10], but this approach requires the use of high-definition cameras to get the benefit of varying resolution. Systems with zoom lenses have some of the advantages of foveated vision, but cannot simultaneously acquire wide angle and high-resolution images.

We follow the first approach (see Figure 5.2) and explore the advantage of foveated vision for object recognition over current approaches, which use equal resolution across the visual field. Although log-polar sensors are a closer match for biology, we note that systems with two cameras in each eye can be advantageous because cameras with standard chips and lenses can be utilized. This makes it possible to equip a humanoid robot with miniature cameras (lipstick size and smaller), which facilitates the mechanical design of the eye and improves its motion capabilities.

FIGURE 5.2   Heads of two humanoid robots. The two robots in the left and central images have foveal cameras mounted above peripheral cameras, and the robot in the right image has foveal cameras on the outer side of the peripheral cameras. See color insert.

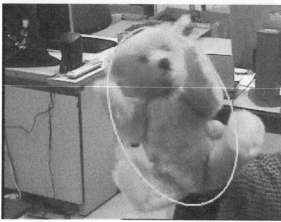

FIGURE 5.3   Simultaneous views from the peripheral and foveal cameras. The high resolution of object image and better localization makes foveal images suitable for recognition (bottom), and the wide field of view from the peripheral camera is suitable for smooth pursuit (top).

In addition, narrow-angle lenses used for foveal cameras in such setups are much less prone to distortion than wide-angle lenses. Systems that use narrow-angle lenses to acquire high-resolution foveal images thus result in higher-quality images than a system with wide-angle lenses, which would necessarily be used if each eye were equipped with only one camera.

There has been a lot of research on oculomotor control and processing of visual information on robots with humanoid vision in the past [20,21,23,26,29]. In this chapter we focus on the utilization of foveated vision for object recognition. Because the fovea covers only a relatively small area of the retina, it is possible to use computationally expensive classification algorithms on the limited portion of the scene on which the fovea is fixated for recognition [13]. This fact has been exploited in References [4,13,33].

5.2 MAINTAINING THE FOVEAL VIEW OF AN OBJECT

In the context of foveated vision, the main task of the eye movement control system is to place objects over the field of view of both foveal cameras so that further analysis and eventually object recognition can be accomplished. Although the focus of the task is to bring the salient areas into the center of the fovea, we propose to use peripheral cameras as the basis for eye movement control. Object tracking in peripheral images is more reliable for smooth pursuit because objects can easily be lost from the foveal views. Two issues need to be clarified when analyzing the foveation setup with two cameras:

1. Given a 3D point that projects onto the center of the foveal image, where is the projection of this same point in the peripheral image? This is the desired position for foveation control in the peripheral image.
2. If a 3D point projects onto the peripheral image away from the ideal position described above, how far is the projection of the point from the center of the foveal image?

5.2.1 CAMERA MODEL

For our theoretical analysis, we model both cameras by a standard pinhole camera model. We denote a 3D point by $M = \begin{bmatrix} X & Y & Z \end{bmatrix}^T$ and a 2D point by $m = \begin{bmatrix} x & y \end{bmatrix}^T$. Let $\tilde{M} = \begin{bmatrix} X & Y & Z & 1 \end{bmatrix}^T$ and $\tilde{m} = \begin{bmatrix} x & y & 1 \end{bmatrix}^T$ be the homogeneous coordinates of M and m, respectively. The relationship between a 3D point M and its projection m is then given by [37]

$$s\tilde{m} = A \begin{bmatrix} R & t \end{bmatrix} \tilde{M} \tag{5.1}$$

where s is an arbitrary scale factor, R and t are the extrinsic parameters denoting the rotation and translation that relate the world coordinate system to the camera coordinate system, and A is the intrinsic matrix

$$A = \begin{bmatrix} \alpha & \gamma & x_0 \\ 0 & \beta & y_0 \\ 0 & 0 & 1 \end{bmatrix} \tag{5.2}$$

α and β are the scale factors, γ is the parameter describing the skewness of the two image axes, and (x_0, y_0) is the principal point.

In the following we assume without loss of generality that the origin of the image coordinate system coincides with the principal point (x_0, y_0), thus $x_0 = y_0 = 0$. Note that on real cameras the principal point does not coincide with the image center in pixel coordinates. However, because the distortion is smallest around the principal point and because making this assumption significantly simplifies the equations, it makes sense to attempt to bring the point of interest to the position that projects onto the principal point of the foveal camera and not onto the precise image center. For standard video cameras with resolution of 640 × 480 pixels, the distance of the principal point from the image center in pixel coordinates is usually less than 10 pixels.

The pinhole camera model (5.1) does not take into account the effects of lens distortion. Such an assumption is justified for foveal cameras, which are equipped with lenses with relatively long focal lengths. Such lenses do not exhibit significant distortion effects. Conversely, to achieve a wide field of view, peripheral cameras typically use lenses with shorter focal lengths, which are significantly more prone to distortion. However, the distortion can be partially corrected in a preprocessing step using a suitable distortion correction procedure, for example, the one described in Reference [37]. Equation (5.1) is valid for the distortion-corrected pixels and we therefore do not consider distortion in our analysis.

5.2.2 TRANSFORMATION OF A PRINCIPAL POINT FROM FOVEAL TO PERIPHERAL IMAGES

We denote by A_f, R_f, t_f and A_p, R_p, t_p the intrinsic and extrinsic parameters of the foveal and peripheral cameras, respectively. Let's now assume that the world coordinate system is aligned with the coordinate system of the foveal camera. In this case we have $R_f = I$, where I is the identity matrix, and $t_f = 0$. Let $\hat{t}$ be the position of the origin of the peripheral coordinate system expressed in the foveal coordinate system and let $\hat{R}$ be the rotation matrix that rotates the basis vectors of the peripheral coordinate system into the basis vectors of the foveal coordinate system. We then have

$$R_p M + t_p = \hat{R}(M - \hat{t}) \tag{5.3}$$

The projections of a 3D point M onto the foveal and peripheral image are then given by

$$x_f = \frac{\alpha_f X + \gamma_f Y}{Z} \tag{5.4}$$

$$y_f = \frac{\beta_f Y}{Z} \tag{5.5}$$

and

$$x_p = \frac{\alpha_p r_1 \cdot (M - \hat{t}) + \gamma_p r_2 \cdot (M - \hat{t})}{r_3 \cdot (M - \hat{t})} \tag{5.6}$$

$$y_p = \frac{\beta_p r_2 \cdot (M - \hat{t})}{r_3 \cdot (M - \hat{t})} \tag{5.7}$$

where r_1, r_2, and r_3 are the rows of the rotation matrix $\hat{R} = \begin{bmatrix} r_1^T & r_2^T & r_3^T \end{bmatrix}^T$. M projects onto the principal point in the fovea if $x_f = y_f = 0$. Assuming that M is in front of the camera, hence $Z > 0$, we obtain from Equations (5.4) and (5.5) that $X = Y = 0$, which means that the point must lie on the optical axis of the foveal camera. Inserting this into Equations (5.6) and (5.7), we obtain the following expression for the desired position $(\hat{x}_p, \hat{y}_p)$ in the peripheral image that results in the projection onto the principal point in the foveal image,

$$\hat{x}_p = \frac{\alpha_p r_1 \cdot t + \gamma_p r_2 \cdot t - (\alpha_p r_{13} + \gamma_p r_{23})Z}{r_3 \cdot t - r_{33}Z} \tag{5.8}$$

$$\hat{y}_p = \frac{\beta_p r_2 \cdot t - \beta_p r_{23} Z}{r_3 t - r_{33} Z} \qquad (5.9)$$

where $\begin{bmatrix} r_{13} & r_{23} & r_{33} \end{bmatrix}^T$ is the third column of $\hat{R}$. Note that the desired position in the periphery is independent of the intrinsic parameters of the foveal camera. It depends, however, on the distance of the point of interest from the eye setup.

5.2.3 DISPLACEMENT FROM THE DESIRED POSITION IN PERIPHERAL AND FOVEAL IMAGES

Let's assume now that the 3D point of interest M projects onto a pixel away from the principal point in the foveal image by displacement (D_x, D_y). From (5.4) and (5.5) we have

$$D_x = \frac{\alpha_f X + \gamma_f Y}{Z} \qquad (5.10)$$

$$D_y = \frac{\beta_f Y}{Z} \qquad (5.11)$$

thus

$$X = \frac{D_x - \gamma_f D_y / \beta_f}{\alpha_f} Z \qquad (5.12)$$

$$Y = \frac{D_y}{\beta_f} Z \qquad (5.13)$$

The 3D point $\begin{bmatrix} 0 & 0 & Z \end{bmatrix}^T$ is the point on the optical axis that is closest to M. It projects onto $(\hat{x}_p, \hat{y}_p)$ in the peripheral view. We define (d_x, d_y) to be the displacement of the projection of M from this point in the peripheral view and we would like to express (D_x, D_y) in terms of (d_x, d_y). We obtain the following relationship,

$$s \begin{bmatrix} \hat{x}_p + d_x \\ \hat{y}_p + d_y \\ 1 \end{bmatrix} = A_p \begin{bmatrix} (r_{11} (D_x - \gamma_f D_y/\beta_f)/\alpha_f + r_{12} D_y/\beta_f + r_{13}) Z - r_1 \cdot t \\ (r_{21} (D_x - \gamma_f D_y/\beta_f)/\alpha_f + r_{22} D_y/\beta_f + r_{23}) Z - r_2 \cdot t \\ (r_{31} (D_x - \gamma_f D_y/\beta_f)/\alpha_f + r_{32} D_y/\beta_f + r_{33}) Z - r_3 \cdot t \end{bmatrix} \qquad (5.14)$$

By subtracting (5.8) and (5.9) from (5.14), we obtain a rather complex expression for the error in the periphery (d_x, d_y) in terms of the error in the fovea (D_x, D_y). This expression also depends on the distance Z of the point of interest from the camera setup. It can be simplified by making some reasonable assumptions about the foveation setup. Usually we construct a foveated vision system with two cameras in such a way that the optical axes of the peripheral and foveal camera are parallel (as in Figure 5.2). The cradles of both cameras must be built with sufficient precision for this purpose. In this case we have $r_{31} = r_{32} = r_{13} = r_{23} = 0$, $r_{33} = 1$ and Equation

(5.14) becomes

$$s \begin{bmatrix} \hat{x}_p + d_x \\ \hat{y}_p + d_y \\ 1 \end{bmatrix} = A_p \begin{bmatrix} \left(r_{11} \left(D_x - \gamma_f D_y/\beta_f \right) /\alpha_f + r_{12} D_y/\beta_f \right) Z - r_1 \cdot t \\ \left(r_{21} \left(D_x - \gamma_f D_y/\beta_f \right) /\alpha_f + r_{22} D_y/\beta_f \right) Z - r_2 \cdot t \\ Z - t_z \end{bmatrix}$$

(5.15)

where $t = \begin{bmatrix} t_x & t_y & t_z \end{bmatrix}^T$. The denominator in (5.8) and (5.9) coincides with the third component in Equation (5.15), thus subtracting (5.8) and (5.9) from (5.15) results in

$$d_x = \frac{Z}{Z - t_z} \left(\frac{\alpha_p r_{11} + \gamma_p r_{21}}{\alpha_f} D_x + \frac{-r_{11}\alpha_p \gamma_f + r_{12}\alpha_p^2 - r_{21}\gamma_p \gamma_f + r_{22}\gamma_p^2}{\alpha_f \beta_f} D_y \right)$$

$$d_y = \frac{Z}{Z - t_z} \left(\frac{r_{21}\beta_p}{\alpha_f} D_x + \frac{\beta_p \left(-r_{21}\gamma_f/\alpha_f + r_{22} \right)}{\beta_f} D_y \right)$$

It is easy to invert this system to obtain the expression for the error in the fovea in terms of the error in the periphery and distance Z. We omit the details here and only present results with further simplifying assumptions. Let's assume that both cameras are completely aligned; that is, $r_{21} = r_{12} = 0$ and $r_{11} = r_{22} = 1$ (no rotation around the optical axis). In this case we can calculate a simpler relationship between the error displacements in the foveal and peripheral images:

$$D_x = \frac{Z - t_z}{Z} \cdot \frac{\alpha_f}{\alpha_p} \left(d_x + \frac{\alpha_p \gamma_f - \gamma_p \alpha_f}{\alpha_f \beta_f} d_y \right)$$

(5.16)

$$D_y = \frac{Z - t_z}{Z} \cdot \frac{\beta_f}{\beta_p} d_y$$

(5.17)

The skew parameters γ_p and γ_f are normally much smaller than $\alpha_p, \alpha_f, \beta_p$, and β_f. Similarly, the displacement of the camera t_z is usually much smaller than the distance Z of the point of interest from the camera. Note that it is difficult to achieve $t_z = 0$ on a practical camera system because it is not easy to determine the exact positions of the projection centers and to mount the cameras accordingly. For totally aligned cameras we finally obtain the following approximation,

$$D_x \approx \frac{\alpha_f}{\alpha_p} d_x$$

(5.18)

$$D_y \approx \frac{\beta_f}{\beta_p} d_y$$

(5.19)

This means that the distance from the desired displacement in the peripheral image is scaled in the fovea by the ratio of focal lengths. This approximation is exact for perfect pinhole cameras ($\gamma_p = \gamma_f = 0$) with precisely aligned coordinate systems; that is, $R = I$ and $t_z = 0$. Because the focal length of the foveal camera is always greater than the focal length of the peripheral camera (i.e., $\alpha_p, \beta_p < \alpha_f, \beta_f$), the deviation from the principal point in the fovea is greater than the deviation from the ideal position in the peripheral image. This is, of course, an expected result.

5.2.4 ANALYSIS OF FOVEATED VISION SYSTEMS WITH TWO CAMERAS

Making the same assumptions as when calculating (5.15), we obtain from Equations (5.8) and (5.9) the following expression for the ideal position in the peripheral image,

$$\hat{x}_p = \frac{\alpha_p r_1 \cdot t + \gamma_p r_2 \cdot t}{t_z - Z} \approx -\frac{\alpha_p r_1 \cdot t}{Z} \tag{5.20}$$

$$\hat{y}_p = \frac{\beta_p r_2 \cdot t}{t_z - Z} \approx -\frac{\beta_p r_2 \cdot t}{Z} \tag{5.21}$$

We can again neglect the influence of γ_p, which is always significantly smaller than α_p and β_p. Note, however, it is important that the cradles are built precisely and that the optical axes are aligned accurately because in Equations (5.8) and (5.9) the zero terms r_{13} and r_{23} are multiplied by Z, which is usually large. Hence if the system is not built precisely, the above approximations are not valid. The above equation system can be further simplified by assuming totally aligned pinhole cameras ($r_{21} = r_{12} = 0$ and $r_{11} = r_{22} = 1$, $t_z = 0$, $\gamma_p = 0$), which results in

$$\hat{x}_p = -\frac{\alpha_p t_x}{Z} \tag{5.22}$$

$$\hat{y}_p = -\frac{\beta_p t_y}{Z} \tag{5.23}$$

In our foveation setup, the peripheral cameras are equipped with 3 mm lenses and with CCD chips of size 6.6×4.4m, and the foveal cameras are equipped with 12 mm lenses and with CCD chips of size 3.3×2.2 m. The distance between them is about 25 mm along the y-axis ($t_x \approx 0$, $t_y \approx 25$). Theoretically, the scaling factors of such cameras are $\alpha_p = \beta_p \approx 290.9$ and $\alpha_f = \beta_f \approx 1,306.8$ when the cameras are calibrated for images of size 640×480. Figure 5.4 shows the variation of $\hat{y}_p$ with respect to Z under these assumptions and proves that our system indeed exhibits such characteristics. For $Z = 1$ m, the desired position in the peripheral image is given by $\hat{x}_p = 0$ and $\hat{y}_p = -290.9 \times 25/1000 = -7.3$ pixels. For objects farther away, $\hat{y}_p$ tends to zero. From Equation (5.23) it follows that the necessary displacement doubles to -14.6 when $Z \approx 498$ mm. Hence, if we fix Z to 1 m and observe objects more than 0.5 m away from the camera, the systematic error in the peripheral images will be less than 7.3 pixels. Equation (5.19) tells us that the displacement from the central position in the foveal view will be at most $1,306.8/290.9 \times 7.3 \approx 32.8$ pixels, hence we are still relatively close to the principal point in the foveal image. Note that fixing the distance Z is equivalent to replacing the perspective projection with the orthographic projection.

5.3   CLOSED-LOOP FOVEATION CONTROL

Gaskett, Ude, and Cheng [12] developed a system for eye movement control that is appropriate for keeping the object in the center of foveal views. It has the appealing property of enhancing the appearance of a humanoid robot through mimicking some aspects of human movement: human eyes follow object movement, but without head (and possibly body) movement have a limited range; thus, the robot's control system

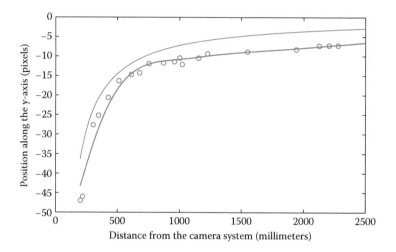

FIGURE 5.4 Top curve: $\hat{y}_p$ with respect to the distance of the object from the camera system as calculated by Equation (5.23) (totally aligned, ideal pinhole cameras, $\hat{R} = I$, $t_x = t_z = 0$, $t_y = 25$, $\alpha_p = \beta_p = 290.9$). Bottom curve: $\hat{y}_p$ determined experimentally by placing the object manually at various distances so that it projects on the center of the foveal image. At each such configuration we measured the object's position in the peripheral image and its distance from the eye (using stereo vision). The circles show these measurements.

supports its eye movements through head (and body) movements. Altogether, the system developed in Reference [12] uses 10 degrees-of-freedom (4 for the eyes + 3 for the head + 3 for the torso) to maintain the view of the object.

The robot's primary mechanism for maintaining the view of the object of interest is eye movement: the control system continuously alters the pan and tilt of each eye to keep the object near the center of the corresponding view. Independent eye motion is acceptable when the object is being tracked properly in both peripheral views, but looks rather unnatural when one eye loses its view of the object while the other eye continues to roam. A possible solution is to introduce a gentle cross-coupling between a camera's view and the control of the other eye. Thus, when a camera's view of the target is lost, its corresponding eye continues to move, fairly slowly, under the influence of the other camera's view. As well as appearing natural, such eye movements improve the likelihood of relocating the object.

The complete control system can be realized as a network of PD controllers expressing the assistive relationships and is given in Reference [12]. The PD controllers are based on simplified mappings between visual coordinates and joint angles rather than on a full kinematic model. Such mappings are sufficient because the system is closed-loop and can make corrective movements to converge toward the desired configuration. Importantly, the developed controller does not rely on exact knowledge of the robot's kinematics and therefore does not deteriorate with the joint and valve wear and tear and maintenance activities.

5.4 LEARNING AND RECOGNIZING OBJECTS IN FOVEAL VIEWS

Our aim is to show that by using foveated vision we can improve the performance of object recognition. For testing purposes we developed a view-based approach that learns a model of the object's appearance in a two-dimensional image under different poses. View-based approaches are popular due to their conceptual simplicity and sometimes impressive recognition performance [6]. We combined probabilistic tracking [34], Gabor jets [36], and multiclass support vector machines [9] to realize our system. In the following we briefly describe the main components.

5.4.1 NORMALIZATION THROUGH AFFINE WARPING

In view-based systems objects are represented by a number of 2D images (or features extracted from them) acquired from different viewpoints. At recognition time these model images are compared to a new test image acquired by the robot. Both a humanoid robot and objects can move in space, therefore objects appear in images at different positions, orientations, and scales. It is obviously not feasible to learn all possible views due to time and memory limitations. The number of required views can, however, be reduced by normalizing the subimages that contain the objects of interest to an image of fixed size.

Such a reduction can be achieved by utilizing the results of an object detection and tracking system. Our tracker can estimate the shape of the observed object using second-order statistics of pixels that are probabilistically classified as "blob pixels" [34]. From the second-order statistics, we can estimate the planar object orientation and the extent of the object along its major and minor axes. In other words, we can estimate the ellipse enclosing the object pixels. As the lengths of both axes can differ significantly, each object image is normalized along the principal axis directions instead of image coordinate axes and we apply a different scaling factor along each of these directions. By aligning the object's axes with the coordinate axes, we also achieve invariance against planar rotations, thus reducing the number of views that need to be stored to represent an object. The results of the described normalization process are shown in Figures 5.5 and 5.6.

5.4.2 GABOR JETS

Early view-based approaches used raw grayscale images as input to the selected classifier, for example, principal component analysis [32]. This kind of approach turned out to be fairly successful as long as the amount of noise in the images remained small and the illumination conditions did not change. To achieve robustness against

FIGURE 5.5   Example images of 10 objects. Scaling and planar rotations are accounted for by affine warping using the results of visual tracking.

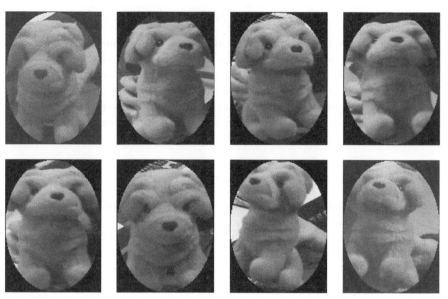

FIGURE 5.6   Training images for one of the objects used in statistical experiments. To take care of rotations in depth, we must collect a sufficient amount of typical viewpoints.

brightness changes, it is necessary to compute an improved, illumination-insensitive characterization of the local image structure. More advanced recognition systems therefore apply a bank of illumination-insensitive filters before starting the recognition process. We follow the biologically motivated approach of Wiskott et al. [36], who proposed applying a bank of Gabor filters to the incoming foveal images. Gabor filters are known to be good edge detectors and are therefore robust against varying brightness. They have limited support both in space and frequency domains and have a certain amount of robustness against translation, distortion, rotation, and scaling [36].

Complex Gabor kernels are defined by

$$\Phi_{\mu,\nu}(\boldsymbol{x}) = \frac{\|\boldsymbol{k}_{\mu,\nu}\|^2}{\sigma^2} \cdot \exp\left(-\frac{\|\boldsymbol{k}_{\mu,\nu}\|^2\|\boldsymbol{x}\|^2}{2\sigma^2}\right) \cdot \left(\exp\left(i\boldsymbol{k}_{\mu,\nu}^T\boldsymbol{x}\right) - \exp\left(-\frac{\sigma^2}{2}\right)\right)$$

(5.24)

where $\boldsymbol{k}_{\mu,\nu} = k_\nu[\cos(\phi_\mu), \ \sin(\phi_\mu)]^T$. The Gabor jet at pixel $\boldsymbol{x}$ is defined as a set of complex coefficients $\{J_j^{\boldsymbol{x}}\}$ obtained by convolving the image with a number of Gabor kernels at this pixel. Gabor kernels are selected so that they sample a number of different wavelengths k_ν and orientations ϕ_μ. Wiskott et al. [36] proposed using $k_\nu = 2^{-((\nu+2)/2)}$, $\nu = 0, \ldots, 4$, and $\phi_\mu = \mu(\pi/8)$, $\mu = 0, \ldots, 7$, but this depends both on the size of the incoming images and the image structure. They showed that the similarity between the jets can be measured by

$$S\left(\{J_i^{\boldsymbol{x}}\}, \{J_i^{\boldsymbol{y}}\}\right) = \frac{\boldsymbol{a}_{\boldsymbol{x}}^T * \boldsymbol{a}_{\boldsymbol{y}}}{\|\boldsymbol{a}_{\boldsymbol{x}}\|\|\boldsymbol{a}_{\boldsymbol{y}}\|}$$

(5.25)

where $a_x = [|J_1^x|, \ldots, |J_s^x|]^T$ and s is the number of complex Gabor kernels. This is based on the fact that the magnitudes of complex coefficients vary slowly with the position of the jet in the image.

We use Gabor jets to generate feature vectors for recognition. To reduce the dimensionality of these feature vectors, we do not use the jets calculated at every pixel. Ideally, one would calculate the jets only at important local features. However, it is very difficult to extract local features in a stable manner. Instead we decided to build the feature vectors from Gabor jets positioned on a regular grid of pixels. Normalized jets $\{a_j^x / \|a^x\|\}_{j=1}^n$ calculated on this grid and belonging to the ellipse enclosing the object as in Figure 5.6 were finally utilized to build feature vectors.

It is important to note that we first scale the object images to a fixed size and then apply Gabor filters. In this way we ensure that the size of the local structure in the acquired images does not change and consequently we do not need to change the frequencies k_v of the applied filters with respect to the image size of the object in foveal images.

5.4.3 DATA ACQUISITION

To learn a viewpoint-independent model, we must show the objects from the database to a humanoid robot from all relevant viewing directions. Turntables are often used to collect images from a sequence of regularly spaced viewpoints. However, this solution is not practical in humanoid robotics, where seamless learning is very important to ensure autonomy. We therefore explored whether it is possible to learn 3D descriptions reliably from images collected while a human teacher moves the object in front of the robot. Using the previously described eye movement control system, the robot tracks the object using the method described in Reference [34], acquires its foveal images from many different viewpoints, and extracts feature vectors. These data are stored to compute a suitable classifier.

5.4.4 RECOGNITION WITH GABOR KERNELS AND SUPPORT VECTOR MACHINES

We employed support vector machines (SVMs) for classification because they deliver high performance in real-world applications. To distinguish between two different classes, a support vector machine draws a separating hyperplane between training datapoints from both classes. The hyperplane is optimal in the sense that it separates the largest fraction of points from each class, while maximizing the distance from either class to the hyperplane. First approaches that utilized SVMs for object recognition used the most standard form of SVMs, which can solve two-class classification problems only. A binary tree strategy [14,24] was proposed to solve the multiclass problem. Although this approach provides a simple and powerful classification framework, it cannot capture correlations between the different classes because it breaks a multiclass problem into multiple independent binary problems. In our work we use the generalization of SVMs to multiclass problems as described in Reference [9].

To design a suitable kernel for classifying feature vectors consisting of Gabor jets, we use the similarity measure between Gabor jets (5.25) as a starting point. Let X_G be the set of all grid points within two normalized images at which Gabor jets are calculated and let $J_{\mathbf{X}_G}$ and $L_{\mathbf{X}_G}$ be the Gabor jets calculated in two different images, but on the same grid

$$K_G(J_{\mathbf{X}_G}, L_{\mathbf{X}_G}) = \exp\left(-\rho_1 \frac{1}{M} \sum_{x \in \mathbf{X}_G} \left(1 - \frac{a_x^T * b_x^T}{\|a_x\| \|b_x\|}\right)\right) \qquad (5.26)$$

where M is the number of grid points in $\mathbf{X}_G$. This function satisfies the so-called Mercer's condition [7] and can thus be used for support vector learning and classification.

5.5   EXPERIMENTAL RESULTS

We used a set of 10 objects to test the performance of the recognition system on a humanoid robot (6 teddy bears, 2 toy dogs, a coffee mug, and a face; see Figure 5.5). For each object we recorded two or more movies using a video stream coming from DB's foveal cameras. The cameras were controlled as described in Section 5.2. In each of the recording sessions the experimenter attempted to show one of the objects to the robot from all relevant viewing directions. One movie per object was used to construct the SVM classifier, and one of the other movies served as input to test the classifiers. Each movie was one minute long and we used at most 4 images per second for training. Slightly more than the first 10 seconds of the movies was needed to initialize the tracker, thus we had at most 208 training images per object. For testing we used 10 images per second, which resulted in 487 test images per object. Except for the results of Table 5.4, all the percentages presented here were calculated using the classification results obtained on 4,870 test images. Two types of classifiers were used to test the performance of foveated recognition. The first was a multiclass support vector machine based on kernel function K_G from Equation (5.26). It is denoted as SVM in Tables 5.1 - 5.4. Gabor jets were calculated at eight different orientations and five different scales and the grid size was 5 pixels in both directions. The filters were scaled appropriately when using lower resolution images. The second classifier was the nearest neighbor classifier (NNC) that uses the similarity measure (5.25)—summed over all grid points—to calculate the nearest neighbor based with respect to the extracted Gabor jets.

The results in Tables 5.1 to 5.3 demonstrate that foveation is very useful for recognition. The classification results clearly become worse with the decreasing resolution.

TABLE 5.1
Classification Rate[a]

Training Views per Object	SVM (%)	NNC (%)
208	97.6	95.9
104	96.7	93.7
52	95.1	91.5
26	91.9	86.7

[a] Image Resolution 120 × 160 pixels.

TABLE 5.2
Classification Rate[a]

Training Views per Object	SVM (%)	NNC (%)
208	94.2	89.3
104	92.4	87.3
52	90.7	84.4
26	86.7	79.2

[a] Image Resolution 60 × 80 Pixels.

TABLE 5.3
Correct Classification Rate[a]

Training Views per Object	SVM (%)	NNC (%)
208	91.0	84.7
104	87.2	81.5
52	82.4	77.8
26	77.1	72.1

[a] Image Resolution 30 × 40 Pixels.

Our results also show that we can collect enough training data even without using accurate turntables to systematically collect the images.

We also tested the performance of the system on data captured under changed lighting conditions (see Figure 5.7). We used two objects (the last teddy bear and the toy dog from Figure 5.5). For classification we used the same SVMs as in Tables 5.1 to 5.3. The performance decreased slightly on darker images; however, the results show that the method still performs well in such conditions. This is due to the properties of Gabor jets and due to the normalization of jets provided by the similarity function (5.25). Our experiments showed that the classification rate drops significantly if one of the standard kernel functions (e. g., a linear kernel) is used for the support vector learning.

TABLE 5.4
Correct classification rate for images acquired under varying lighting conditions (see Figure 5.7). Only two objects were tested in this case (the database still contained ten objects) and SVMs calculated based on 208 views per training objects were used.

Image Resolution	Normal (%)	Dark (%)	Very Dark
120 × 160	99.5	97.7	97.9
60 × 80	96.7	93.5	95.0
30 × 40	93.6	89.3	88.2

FIGURE 5.7   Images acquired under different lighting conditions.

5.6 CONCLUSIONS

In this chapter we presented the design of foveated vision systems on humanoid robots. We showed that by utilizing a humanoid vision system that mimics the foveated structure of a human eye, supplemented by appropriate oculomotor control algorithms and computational models of vision processing, we can significantly improve object recognition compared to standard robot vision systems. However, it is still not possible to achieve recognition performance and robustness similar to humans. Although recent advances in the neuroscience of vision enabled researchers to collect vast amounts of new data, advanced computational models of brain processes that result in object recognition are still not available. A tighter integration between the models of activity in certain areas of the brain and functional models of vision processing is needed. Until this happens, high-performance technical systems will continue to rely on statistical approaches that do not necessarily contribute to our understanding of computational processes in the brain.

Advances in basic robot technologies are also necessary. Robust hardware that allows the implementation of long-term developmental processes is needed. Current robotic systems are still too fragile to replicate the process of human development. Like many other cognitive processes, human performance at object recognition is also the result of a developmental process that takes many years [22]. It is unlikely that human levels of performance can be achieved without the ability to replicate at least some of the developmental processes that lead to it.

REFERENCES

1. Asfour, T., Welke, K., Azad, P., Ude, A., and Dillmann, R. The Karslruhe humanoid head. In *8th IEEE-RAS International Conference on Humanoid Robots*, Daejeon, Korea, pp. 447–453, 2008.
2. Atkeson, C.G., Hale, J., Pollick, F., Riley, M., Kotosaka, S., Schaal, S., Shibata, T., Tevatia, G., Ude, A., Vijayakumar, S., and Kawato, M. Using humanoid robots to study human behavior. *IEEE Intell. Syst.* **15**(4):46–56, July/August 2000.
3. Biederman, I. Recognition-by-components: A theory of human image understanding. *Psychol. Rev.* **94**(2):115–147, 1987.
4. Björkman, M. and Kragic, D. Combination of foveal and peripheral vision for object recognition and pose estimation. In *IEEE International Conference on Robotics and Automation*, New Orleans, pp. 5135–5140, 2004.

5. Breazeal, C., Edsinger, A., Fitzpatrick, P., and Scassellati, B. Social constraints on animate vision. *IEEE Trans. Syst. Man Cyberne. A: Syst. Humans* **31**(5):443–452, July/August 2001.

6. Bülthoff, H.H., Wallraven, C., and Graf, A. View-based dynamic object recognition based on human perception. In *16th International Conference on Pattern Recognition*, Quebec, pp. 768–776, 2002.

7. Burges, C.J.C. A tutorial on support vector machines for pattern recognition. *Data Min. Knowl. Discov.*, **2**(2):121–167, 1998.

8. Cheng, G., Hyon, S.-H., Morimoto, J., Ude, A., Hale, J.G., Colvin, G., Scroggin, W., and Jacobsen, S.C. CB: A humanoid research platform for exploring neuroscience. *Adv. Robot.* **21**(10):1097–1114, 2007.

9. Crammer, K. and Singer, Y. On the algorithmic implementation of multiclass kernel-based vector machines. *J. Mach. Learn. Res.* **2**:265–292, 2001.

10. Engel, G., Greve, D.N., Lubin, J.M., and Schwartz, E.L. Space-variant active vision and visually guided robotics: Design and construction of a high-peformance miniature vehicle. In *12th IAPR International Conference on Pattern Recognition*, Jerusalem, Israel, pp. 487–490, 1994.

11. Fischer, B. and Ramsperger, E. Human express saccades: Extremely short reaction times of goal directed eye movements. *Exper. Brain Res.* **57**:191–195, 1984.

12. Gaskett, C., Ude, A., and Cheng, G. Hand-eye coordination through endpoint closed-loop and learned endpoint open-loop visual servo control. *Int. J. Human. Robot.* **2**(2):203–224, 2005.

13. Gould, S., Arfvidsson, J., Kaehler, A., Sapp, B., Messner, M., Bradski, G., Baumstarck, P., Chung, S., and Ng, A.Y. Peripheral-foveal vision for real-time object recognition and tracking in video. In *20th International Joint Conference on Artificial Intelligence*, Hyderabad, pp. 2115–2121, 2007.

14. Guo, G., Li, S.Z., and Chan, K.L. Support vector machines for face recognition. *Image Vis. Comput.* **19**(9-10):631–638, 2001.

15. Hoffmann, M.B. and Bach, M. The distinction between eye and object motion is reflected by the motion-onset visual evoked potential. *Exper. Brain Res.* **144**:141–151, 2002.

16. Hubel, D.H. *Eye, Brain, and Vision.* 2nd edition, NewYork: W.H. Freeman, 1995.

17. Kozima, H. and Yano, H. A robot that learns to communicate with human caregivers. In *International Workshop on Epigenetic Robotics*, Lund, Sweden, 2001.

18. Larson, A.M. and Loschky, L.C. The contributions of central versus peripheral vision to scene gist recognition. *J. Vis.* **9**(10):1–16, 2009.

19. Levy, I., Hasson, U., Avidan, G., Hendler, T., and Malach, R. Center-periphery organization of human object areas. *Nature Neurosci.* **4**:533–539, 2001.

20. Manzotti, R., Gasteratos, A., Metta, G., and Sandini, G. Disparity estimation on log-polar images and vergence control. *Comput. Vis. Image Understand.* **83**(2):97–117, 2001.

21. Metta, G., Gasteratos, A., and Sandini, G. Learning to track colored objects with log-polar vision. *Mechatronics* **14**:989–1006, 2004.

22. Nishimura, M., Scherf, S., and Behrmann, M. Development of object recognition in humans. *F1000 - Biol. Rep.* **1**:1–4, 2009.

23. Panerai, F., Metta, G., and Sandini, G. Visuo-inertial stabilization in space-variant binocular systems. *Robot. Auton. Syst.* **30**(1–2):195–214, 2000.

24. Pontil, M. and Verri, A. Support vector machines for 3D object recognition. *IEEE Trans. Patt. Anal. Mach. Intell.* **20**(6):637–646, 1998.

25. Preissig, J.J. and Tarr, M.J. Visual object recognition: Do we know more now than we did 20 years ago? *Ann. Rev. Psychol.* **58**:75–96, 2007.

26. Rougeaux, S. and Kuniyoshi, Y. Robust tracking by a humanoid vision system. In *IAPR First International Workshop on Humanoid and Friendly Robotics*, Tsukuba, Japan, 1998.

27. Sandini, G. and Metta, G. Retina-like sensors: motivations, technology and applications. In F.G. Barth, J.A.C. Humphrey, and T.W. Secomb (Eds.), *Sensors and Sensing in Biology and Engineering*. Wien-New York: Springer-Verlag, 2003.

28. Scassellati, B. Eye finding via face detection for a foveated, active vision system. In *Fifteenth National Conference on Artificial Intelligence*, Madison, Wisconsin, pp. 969–976, 1998.

29. Shibata, T., Vijayakumar, S., Jörg Conradt, J., and Schaal, S. Biomimetic oculomotor control. *Adapt. Behav.* **9**(3/4):189–208, 2001.

30. Tarr, M.J. and Bülthoff, H.H. Image-based object recognition in man, monkey, and machine. *Cognition* **67**(1–2):1–20, 1998.

31. Thier, P. and Ilg, U.J. The neural basis of smooth-pursuit eye movements. *Curr. Opin. Neurobiol.* **15**:645–652, 2005.

32. Turk, M. and Pentland, A. Eigenfaces for recognition. *J. Cogn. Neurosci.* **3**(1):71–86, 1991.

33. Ude, A., Atkeson, C.G., and Cheng, G. Combining peripheral and foveal humanoid vision to detect, pursue, recognize and act. In *IEEE/RSJ International Conference on Intelligent Robots and Systems*, Las Vegas, pp, 2173–2178, 2003.

34. Ude, A., Shibata, T., and Atkeson, C.G. Real-time visual system for interaction with a humanoid robot. *Robot. Auton. Syst.* **37**(2–3):115–125, 2001.

35. Vera-Díaz, F.A. and Doble, N. The human eye and adaptive optics. In R. K. Tyson (Ed.), *Topics in Adaptive Optics*, pp. 119–150. Rijeka, Croatia: InTech, 2011.

36. Wiskott, L., Fellous, J.-M., Krüger, N., and von der Malsburg, C. Face recognition by elastic bunch graph matching. *IEEE Trans. Patt. Anal. Mach. Intell.* **19**(7):775–779, 1997.

37. Zhang, Z. A flexible new technique for camera calibration. *IEEE Trans. Patt. Anal. Mach. Intell.* **22**(11):1330–1334, 2000.

6 Representation and Control of the Task Space in Humans and Humanoid Robots

Michael Mistry and Stefan Schaal

CONTENTS

6.1 INTRODUCTION

From a robotics perspective, the human body is a complex, high degree-of-freedom motor system, with a plethora of sensory inputs, motor outputs, and multiple end-effectors. The challenge for the nervous system is to synergistically coordinate motion

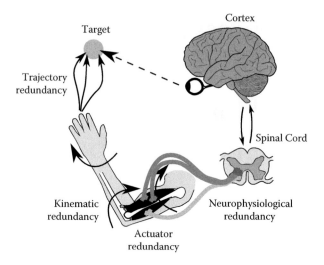

FIGURE 6.1 Even a simple reaching movement requires complex coordination between multiple muscles and joints. The motor control system must cope with compounding layers of redundancy, including *trajectory redundancy* (several possible hand paths, velocities, and accelerations may be used to reach the target), *kinematic redundancy* (a particular hand trajectory may be realized by different possible joint motions), *actuator redundancy* (multiple muscles may actuate a single joint, requiring a choice of how to regulate levels of force and/or stiffness), and finally *neurophysiological redundancy* (a single muscle may be activated by multiple motor neurons, and a single motor neuron may activate a collection of muscles).

and forces from collective muscles, limbs, and joints, to skillfully and efficiently achieve tasks such as reaching, grasping, and walking. The complexity of planning and controlling such skills is daunting, particularly as most of the motions we execute are *redundant* (having an infinite set of possible motor inputs that can achieve the same motor output; Figure 6.1). How the nervous system chooses its particular motor solution, even for the simplest of tasks, is still an active topic in neuroscience research. Roboticists, however, commonly work with redundant systems, and have developed computational methods for reducing complexity. Given a model of a redundant robotic system, with specified end-effector(s), techniques exist for decomposing the robot's kinematics and dynamics into the *task space* (the lower-dimensional subspace of motion and forces directly relevant for task achievement). Orthogonal to the task space is the *null space* (the subspace of task-irrelevant motion and forces). These decomposition methods enable the planning and control of tasks in a simplified, reduced dimensional space, while still allowing for exploitation of redundancy for achieving secondary tasks or absorbing disturbances.

In neuroscience, there has been previous study of how humans plan and coordinate movement within a task space. As early as 1930, Lashley recognized the importance of kinematic redundancy for *motor equivalence* (the ability to achieve the same motor function with multiple but unequal movement) [10]. However, it was Bernstein who first formalized the degrees-of-freedom problem: how does the nervous system cope with the indeterminate mapping from goals to actions [4]? Bernstein postulated that the nervous system is capable of functionally "freezing" or tightly coupling joints, a hypothesis that still holds today, particularly in the study of motor learning. When examining eye movement and gaze, Donder observed a specific and fixed eye torsion for every gaze angle, and thus proposed that the nervous system may have a fixed one-to-one mapping from task variables to joint variables (i.e., Donders law). Limb movements, however, have been shown to violate this relatively simple law [28]. In more recent work, Scholz and Schöner demonstrated via repeated sit-to-stand movements that humans consistently controlled their center of mass, as opposed to task-irrelevant and highly variable hand motion [21]. Lockhar and Ting show how a task-level center-of-mass acceleration can be used to predict synergies of muscle activation in the legs of balancing cats [11]. Todorov and Jordan suggest the *minimum intervention principle* (motor output errors are only corrected when they interfere with task performance) [30].

The goal of this chapter is to bring to light some of the current robotics methodology in task space and redundant control, and to demonstrate their applications toward the modeling of human movement control and redundancy resolution. We first describe, primarily from the robotics point of view, the theory and methodology for representing and controlling task-space motion and forces. We show how these modeling techniques are a useful tool for probing the nervous system's strategies for motor control, learning, and adaptation. We first introduce the notation and methodology for task-space control in robotics, including inverse kinematics and operational space control. Next, we show evidence of task space control and learning from an experiment in human reaching. We propose a controller for such a movement that exploits an internal model of task-space dynamics, while allowing the disturbance to alter redundant motion. Finally we describe and present solutions to some of the challenges of extending these models from single limbs into whole-body domains, where underactuation and external contacts must be considered.

6.1.1 Preliminaries

We assume the robot with n degrees-of-freedom (DOFs) is represented by the configuration vector $\mathbf{q} \in \mathbb{R}^n$. We also have an m DOF task ($m \leq n$), for example, control of an end-effector, represented by the task-space configuration vector $\mathbf{x} = \mathbf{f}(\mathbf{q}) \in \mathbb{R}^m$, and the *task Jacobian* $\mathbf{J}(\mathbf{q})$, as defined by the relations $\mathbf{J}(\mathbf{q}) = \partial \mathbf{f}/\partial \mathbf{q}$ and $\dot{\mathbf{x}} = \mathbf{J}(\mathbf{q})\dot{\mathbf{q}}$. To control motion within the task space, we can compute a configuration velocity to realize some desired task velocity $\dot{\mathbf{x}}_d$ as follows,

$$\dot{\mathbf{q}}_d = \mathbf{J}^{\#}\dot{\mathbf{x}}_d + \left(\mathbf{I} - \mathbf{J}^{\#}\mathbf{J}\right)\xi \tag{6.1}$$

where $\mathbf{J}^{\#}$ is a pseudoinverse of $\mathbf{J}$, $\mathbf{I}$ is the identity matrix, and ξ is an arbitrary vector. The choice of pseudoinverse is not unique. One common choice is the Moore–Penrose pseudoinverse, which will compute a vector $\mathbf{q}$ with minimum possible norm (assuming $\xi = \mathbf{0}$). Computing $\dot{\mathbf{q}}_d$ using (6.1) with any ξ and the MP pseudoinverse will achieve $\dot{\mathbf{x}}_d$ in the task space provided $\mathbf{J}$ is full row rank. Similarly as velocity, the configuration acceleration can be computed as

$$\ddot{\mathbf{q}}_d = \mathbf{J}^{\#} \left(\ddot{\mathbf{x}}_d - \dot{\mathbf{J}}\dot{\mathbf{q}} \right) + \left(\mathbf{I} - \mathbf{J}^{\#}\mathbf{J} \right) \xi \tag{6.2}$$

Once the configuration velocities or accelerations are known, position can be determined by numerical integration; for example, $\mathbf{q}_{d,i+1} = \mathbf{q}_{d,i} + \dot{\mathbf{q}}\Delta T$ (where i is the current time index and ΔT is the time interval). Control input to the robot can then be determined by a feedback controller, such as

$$\mathbf{u} = \mathbf{K}_P \left(\mathbf{q}_d - \mathbf{q} \right) + \mathbf{K}_D \left(\dot{\mathbf{q}}_d - \dot{\mathbf{q}} \right) \tag{6.3}$$

where $\mathbf{K}_P$ and $\mathbf{K}_D$ are position and velocity gain matrices, respectively. This controller, which we call inverse kinematics with position feedback control, is diagrammed in Figure 6.2A.

If we have a model of the dynamics of the system, this knowledge can be exploited for our controller. The rigid-body inverse-dynamics equation of a robot can be written as

$$\mathbf{M}\left(\mathbf{q}\right)\ddot{\mathbf{q}} + \mathbf{h}\left(\mathbf{q}, \dot{\mathbf{q}}\right) = \tau \tag{6.4}$$

where $\mathbf{M}$ is the inertia matrix; $\mathbf{h}$ is the vector of centrifugal, Coriolis, and gravity forces; and τ is the vector of actuator forces/torques. As diagrammed in Figure 6.2B, the inverse dynamics model can be used to compute feedforward control commands, and a feedback controller is still included to account for disturbances or inaccuracies in the dynamics model. The complete inverse dynamics control equation is written as follows,

$$\mathbf{u} = \mathbf{M}\left(\mathbf{q}\right)\ddot{\mathbf{q}}_d + \mathbf{h}\left(\mathbf{q}, \dot{\mathbf{q}}\right) + \mathbf{K}_P \left(\mathbf{q}_d - \mathbf{q} \right) + \mathbf{K}_D \left(\dot{\mathbf{q}}_d - \dot{\mathbf{q}} \right) \tag{6.5}$$

Instead of mapping the desired task-space motion into joint space motion, via inverse kinematics, we can compute the task-space forces directly. In [9], Khatib derives the dynamics of the task (or operational space) as

$$\Lambda\ddot{\mathbf{x}} + \Lambda \left(\mathbf{JM}^{-1}\mathbf{h} - \dot{\mathbf{J}}\dot{\mathbf{q}} \right) = \mathbf{F} \tag{6.6}$$

where $\Lambda = \left(\mathbf{JM}^{-1}\mathbf{J}^{T} \right)^{-1}$ and $\mathbf{F}$ is an external force applied at the end-effector. Then Khatib formulated the operational space control equation, for redundant manipulators $(m < n)$, as

$$\tau = \mathbf{J}^{T}\mathbf{F} + \left(\mathbf{I} - \mathbf{J}^{T}\bar{\mathbf{J}}^{T} \right) \tau_0 \tag{6.7}$$

where $\mathbf{F}$ is defined by (6.6) using a desired task acceleration $\ddot{\mathbf{x}}_d$ in place of $\ddot{\mathbf{x}}$, and $\bar{\mathbf{J}}$ is the inertia weighted pseudoinverse of $\mathbf{J}$:

$$\bar{\mathbf{J}} = \mathbf{M}^{-1}\mathbf{J}^{T} \left(\mathbf{JM}^{-1}\mathbf{J}^{T} \right)^{-1} \tag{6.8}$$

As discussed by Khatib, this generalized inverse is defined to be dynamically consistent with the task: it is the only generalized inverse that results in zero end-effector acceleration for any τ_0. This inverse also solves the equation $\dot{\mathbf{x}} = \mathbf{J}\dot{\mathbf{q}}$ for the joint velocities that minimize the instantaneous kinetic energy of the system. With (6.7) we are able to compensate for task-space dynamics, such that $\ddot{\mathbf{x}} = \ddot{\mathbf{x}}_d$, and the decoupling motion generated by τ_0 from affecting task-space motion. A null space component is typically still included in the controller to resolve any remaining redundancy. A diagram of the operational space controller is shown in Figure 6.2C.

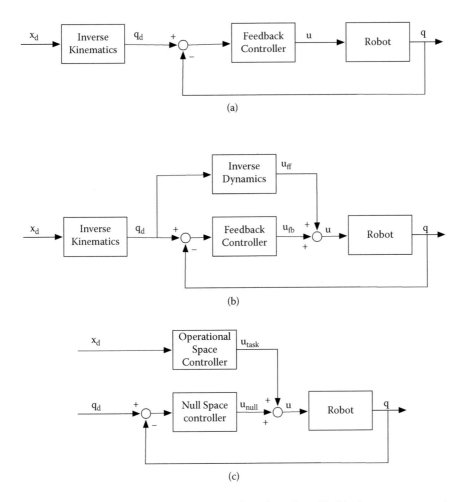

FIGURE 6.2 The three task-space control paradigms investigated in this chapter. x_d represents the desired task, q the robot configuration, and u the robot control input. (A) Position control: the desired task is converted into desired joint positions via inverse kinematics, and subsequently realized by a feedback controller. (B) Inverse dynamics control: feedforward control input u_{ff} is computed via a model of inverse dynamics. Feedback control u_{fb} is used to maintain stability in light of disturbances and modeling errors. (C) Operational space control: the control input required to achieve the task, u_{task}, is directly computed via operational space control. Redundancy is typically resolved using null space control input u_{null}.

6.2   TASK-SPACE CONTROL AND MODELING IN HUMAN MOVEMENT

A primary goal of this chapter is to demonstrate how the previously described tools, developed for the control of robotic systems, may also be used as an insightful model of human movement. Even the simple task of moving your hand to a target (e.g., to manipulate or grasp an object) is highly redundant with an infinite number of possible solutions. To understand which control strategy may be employed by the nervous system, we would like to test the motor control system by altering the dynamics of such a movement and studying how the motor output, particularly the redundant components of the movement, may adapt in response. Motor control researchers have primarily studied the issue of redundancy in human movement only in a kinematic setting, by observing natural human motion and examining patterns that emerge. For example, by examining variance patterns in joint trajectories, certain features or coupled movements may appear consistently in repeated trials. The features that appear consistently over repeated trials can thus be considered as *task relevant*. For example, Scholz and Schöner [21] mapped the features with highest movement variance onto a lower-dimensional manifold (which they coined the *uncontrolled manifold*). However, we would also like to be able to address how these patterns of joint motion and task relevance can potentially change and adapt when learning new skills or new dynamic environments. By studying the problem in a dynamic setting, and potentially altering the dynamics of motion, we can further assess which features of kinematics and/or dynamics become relevant for a specific task such as reaching.

Robot manipulanda have been a common tool in motor control research for altering motor dynamics. In these experiments, robotic manipulanda apply controlled, extraneous forces/torques either at the hand [25], [7], [6], [8] or individual joints [27] while the subjects carry out movement tasks such as point-to-point reaching movements or continuous patterns [6]. However, because of the mechanical constraints of the manipulanda used, these experiments have been limited to two degrees-of-freedom, focusing on only shoulder and elbow joints, and thus not allowing for any spatial redundancy in a movement. With the goal of extending this experimental paradigm to explore a wider variety of movements, including movements in full 3D space using the major seven degrees-of-freedom of the human arm, we have developed an experimental platform using a 7-DOF anthropomorphic exoskeleton (Figure 6.3) [13,14]. The exoskeleton robot allows the subject to make unconstrained 3D movements with all seven major DOFs of the arm. By applying perturbing forces at any or all of these seven joints, we are able to explore how subjects cope with intrinsic (joint space) force fields and resolve redundancy during reaching tasks.

In this section, we first introduce the experimental exoskeleton apparatus. We then discuss our particular reaching movement experiment, and show results from human subjects. We propose a task-space control model to explain the underlying behavior observed in our subjects, and suggest a possible control mechanism used by the nervous system. Finally we draw some conclusions from this work.

6.2.1 EXPERIMENTAL METHODS

The experimental platform is a seven-DOF hydraulically actuated exoskeleton robot arm (Sarcos Master Arm, Sarcos, Inc., Salt Lake City; Figure 6.3). Its anthropomorphic design mimics the major seven DOFs of the human arm,* such that any joint

---

* The major joints of the human arm are labeled: shoulder-flexion–extension (SFE), shoulder-abduction–adduction (SAA), humeral rotation (HR), elbow-flexion–extension (EB), wrist-supination–pronation (WSP), wrist-flexion–extension (WFE), and wrist-abduction–adduction (WAA).

FIGURE 6.3 SARCOS master arm with user.

movement of the user's arm will be approximately reflected by a similar joint movement of the exoskeleton. Conversely, torques applied by an exoskeleton joint will be reflected to the corresponding joint of the user's arm. The exoskeleton's most proximal joint is mounted to a fixed platform, and the user wields the device by holding on to a handle at the most distal joint and by a strapping of the forearm to the equivalent link of the robot just before the elbow. The shoulder remains unconstrained, but is positioned such that the three shoulder rotation axes of the exoskeleton approximately intersect with the human's shoulder joint. Details of the control algorithm for the exoskeleton can be found in the appendix.

We asked healthy, right-handed subjects to make point-to-point reaching movements to a visual target, while wielding the exoskeleton device. In each trial, the right hand begins above the shoulder (as if starting to throw a ball) and finishes with the arm slightly extended, in front of the torso (as if shaking someone's hand). The movement was chosen as such to maximize the spatial extent of the trajectory within the limited workspace of the robot. At the start of each trial, the robot servos the subject's arm to the starting location, and each trial begins with the same joint configuration. The subject is not instructed on how to execute the movement, other than to make the movement as fast as possible, and only use arm motion (not torso, hips, etc.) This way the subject's shoulder position stays approximately at the intersection of the exoskeleton's three shoulder rotation axes, and the subject's and exoskeleton's joint angles will remain in cohesion with minimal discrepancy.

The perturbing force field, when applied, adds joint torques dependent on the velocity of one or several joints. The general form of such external forces is

$$\mathbf{u}_{\text{field}} = \mathbf{A}\dot{\mathbf{q}} \tag{6.9}$$

We investigated two different force fields. Field 1 applies an elbow torque dependent on the sum of shoulder–flexion–extension and shoulder abduction–adduction velocities. The $\mathbf{A}$ matrix of this field takes the form:

$$\mathbf{A}_1 = \begin{bmatrix} 0 & 0 & 0 & 0 & 0 & 0 & 0 \\ 0 & 0 & 0 & 0 & 0 & 0 & 0 \\ 0 & 0 & 0 & 0 & 0 & 0 & 0 \\ 3.0 & 3.0 & 0 & 0 & 0 & 0 & 0 \\ 0 & 0 & 0 & 0 & 0 & 0 & 0 \\ 0 & 0 & 0 & 0 & 0 & 0 & 0 \\ 0 & 0 & 0 & 0 & 0 & 0 & 0 \end{bmatrix} \tag{6.10}$$

Field 2 applies a humeral rotation torque dependent on elbow velocity:

$$\mathbf{A}_2 = \begin{bmatrix} 0 & 0 & 0 & 0 & 0 & 0 & 0 \\ 0 & 0 & 0 & 0 & 0 & 0 & 0 \\ 0 & 0 & 0 & 1.5 & 0 & 0 & 0 \\ 0 & 0 & 0 & 0 & 0 & 0 & 0 \\ 0 & 0 & 0 & 0 & 0 & 0 & 0 \\ 0 & 0 & 0 & 0 & 0 & 0 & 0 \\ 0 & 0 & 0 & 0 & 0 & 0 & 0 \end{bmatrix} \tag{6.11}$$

In experiment 1, each subject completes a block of 70 trials without a force field (null field trials), followed by a block of 70 trials with field 1 turned on. In experiment 2, each subject conducted a block of 70 null trials, followed by 70 trials with force field 2. Experiments 1 and 2 were conducted on different days, and in random order for each subject.

6.2.2 RESULTS

Figure 6.4 (Left) shows the average of the last 10 hand trajectories during one subject's null field block in black and the average of the last 10 hand trajectories during the field 1 trials in gray. On the right, are the averaged joint-space trajectories of shoulder-flexion–extension, shoulder-abduction–adduction, and elbow (SFE, SAA, and EB, respectively) for the same subject during the same trials. Clearly, after force field training, this joint space trajectory is skewed to the right. Average trajectories are found by resampling each trajectory (by applying an antialiasing FIR filter) to a fixed array length, and then computing the mean of each index across all trials.

Figure 6.5 (Left) shows the average of the last 10 hand trajectories during another subject's null field block in black and the average of the last 10 hand trajectories during the field 2 trials in gray. On the right, are the averaged joint-space trajectories of shoulder-abduction–adduction, humeral rotation, and elbow (SAA, HR, and EB, respectively) for the same subject during the same trials.

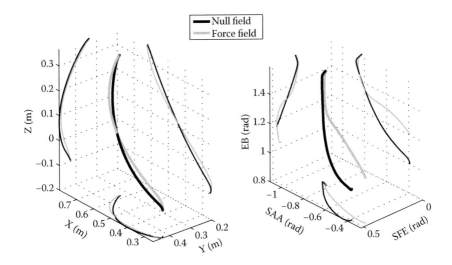

FIGURE 6.4 (Left) Three-dimensional trajectories of the right hand of Subject A during experiment 1. The average of the last 10 null field trials (black) is plotted with the average of the last 10 force field trials (gray). Also plotted as thinner lines are the two-dimensional projections on orthogonal planes. Hand trajectories return to the original null-field trajectories after adapting to the force field. (Right) Average trajectories of the same trials in a three-dimensional projection of joint space (SFE, SAA, and EB). Although the hand trajectory nearly matches the null field trajectory, there is an alteration in joint space.

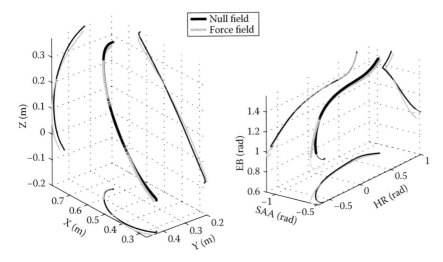

FIGURE 6.5 (Left) Three-dimensional trajectories of the right hand of Subject B during experiment 1. The average of the last 10 null field trials (black) is plotted with the average of the last 10 force field trials (gray). Also plotted as thinner lines are the two-dimensional projections on orthogonal planes. Hand trajectories return to the original null-field trajectories after adapting to the force field. (Right) Average trajectories of the same trials in a three-dimensional projection of joint space (SAA, HR, and EB). Although the hand trajectory nearly matches the null field trajectory, there is an alteration in joint space.

The results demonstrate that the hand trajectory remains invariant after force field adaptation. Interestingly, however, we see no such invariance in joint space trajectories. These results suggest that the motor system may be planning reaching movements in extrinsic (task-space) coordinates and not creating fixed positional trajectories for the joints to follow. Rather, the motor system is able to exploit the redundancy available in the human arm (perhaps to reduce the effects of the force field) while still executing its endpoint plan.

6.2.3 TASK-SPACE CONTROL MODEL

To better understand the underlying process that results in the observed adaptation in our experiment, we propose a model for the motor control of reaching tasks. Our model emphasizes task-space, or operational space control [9], as well as the minimum intervention principle [30]. A disturbance is only corrected for if it interferes with the task, otherwise the nervous system need not be concerned.

First we model the human arm as a 7-DOF rigid body, identical to the form of (6.4). We develop a controller for this arm model that executes a specified task. In this case, our task is to track a 3-DOF desired hand trajectory. The remaining redundancy should be additionally resolved, but in such a way that it does not interfere with task achievement. We use Khatib's formulation of operational space control, which provides a method to decouple those forces applied directly to a task and those in the task's orthogonal null space. Thus, forces can be applied to the null space of a task, such that they do not interfere dynamically with task-space motion.

Given a set of generalized forces $\mathbf{u}$ and a task $\mathbf{x}$ (in this case we assume $\mathbf{u} \in \mathbb{R}^7$ and $\mathbf{x} \in \mathbb{R}^3$), if there is the differential kinematic relationship $\dot{\mathbf{x}} = \mathbf{J}\dot{\mathbf{q}}$, we can separate the generalized forces into their orthogonal task- and null-space components as follows:

$$\mathbf{u} = \mathbf{u}_{\text{task}} + \mathbf{u}_{\text{null}} \tag{6.12}$$

$$= \mathbf{J}^T \bar{\mathbf{J}}^T \mathbf{u} + \left(\mathbf{I} - \mathbf{J}^T \bar{\mathbf{J}}^T\right) \mathbf{u} \tag{6.13}$$

where $\bar{\mathbf{J}}$ is the inertia weighted pseudoinverse of $\mathbf{J}$ as defined in (6.8). Using this methodology, we propose the following control model for a reaching task,

$$\mathbf{u} = \mathbf{J}^T \bar{\mathbf{J}}^T \left(\mathbf{M}\mathbf{J}^+ \left(\ddot{\mathbf{x}}_r - \dot{\mathbf{J}}\dot{\mathbf{q}}\right) + \mathbf{u}_{\text{field}}\right) + \left(\mathbf{I} - \mathbf{J}^T \bar{\mathbf{J}}^T\right) \ddot{\mathbf{q}}_r + \mathbf{h} \tag{6.14}$$

where

$$\ddot{\mathbf{x}}_r = \ddot{\mathbf{x}}_d + \mathbf{K}_{Dx} \left(\dot{\mathbf{x}}_d - \dot{\mathbf{x}}\right) + \mathbf{K}_{Px} \left(\mathbf{x}_d - \mathbf{x}\right) \tag{6.15}$$

$$\ddot{\mathbf{q}}_r = \ddot{\mathbf{q}}_d + \mathbf{K}_{Dq} \left(\dot{\mathbf{q}}_d - \dot{\mathbf{q}}\right) + \mathbf{K}_{Pq} \left(\mathbf{q}_d - \mathbf{q}\right) \tag{6.16}$$

($\mathbf{x}_d, \dot{\mathbf{x}}_d, \ddot{\mathbf{x}}_d$ and $\mathbf{q}_d, \dot{\mathbf{q}}_d, \ddot{\mathbf{q}}_d$ are the desired task and joint trajectories, respectively). Here we choose $\mathbf{K}_{Dx} \gg \mathbf{K}_{Dq}$ and $\mathbf{K}_{Px} \gg \mathbf{K}_{Pq}$ to emphasize the importance of task-space control over joint-space control. Additionally, we only compensate for the force field $\mathbf{u}_{\text{field}}$ in task space. With this model, we assume that the nervous system only learns and compensates for the component of the disturbance affecting hand motion, and allows the disturbance to alter the task-irrelevant null-space motion. Note, in our model we also compensate for Coriolis, centrifugal, and gravity forces in joint space, which assumes the subject has an a priori model of these forces.

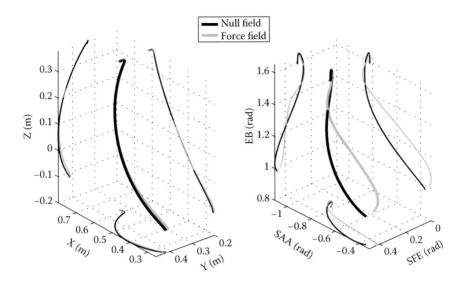

FIGURE 6.6 Simulated results. (Left) Hand trajectories of the rigid-body arm model. A null field trial (black) is plotted together with a force field trial (gray). (Right) Joint space trajectories (SFE, SAA, and EB). The force field perturbs the joint motion similarly as seen in the human data.

We implement our rigid-body arm model and controller using the SL dynamics simulator. For the simulation, we use the same kinematic and estimated inertial parameters of our robot (as found in Section 6.6.2). Ideally we could use inertial parameters that are closer to those of a human arm, but because our controller is compensating for inertial effects, we believe the desired effect should be similar. We tune the gain parameters of (6.14), in order to match our experimental results, but keeping task-space gains at least 20 times higher than joint-space gains. For each experiment, the desired task- and joint-space trajectories are directly taken from the subject's averaged hand and joint-space null-field trajectories. Figure 6.6 shows the simulated result of the condition of Experiment 1 and using our model controller. Qualitatively, the joint space motion under the force field is similar to that found in the actual subject data in Figure 6.4. Figure 6.7 shows the simulated result of Experiment 2. Again, qualitatively, the joint space motion under the force field is similar to that found in Figure 6.5. As an additional test, we dropped the task-space force field compensation in the controller. Figure 6.8 shows the simulated result using force field 1. This is to show that the task-space compensation term is indeed required in order to maintain the desired hand path.

6.2.4 CONCLUSION

With this experiment we demonstrate how a complex exoskeleton robot can be employed for behavioral studies of motor control. Despite the fact that this hydraulic robot cannot compete with the quality of impedance control of very small-scale haptic

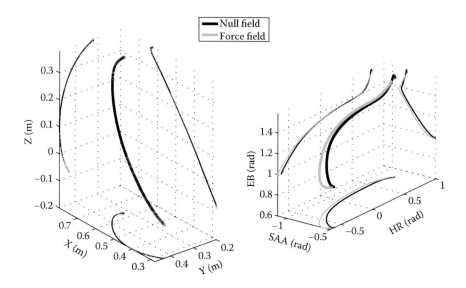

FIGURE 6.7 Simulated results. (Left) Hand trajectories of the rigid-body arm model. A null field trial (black) is plotted together with a force field trial (gray). (Right) Joint space trajectories (SAA, HR, and EB). The force field perturbs the joint motion similarly to that seen in the human data.

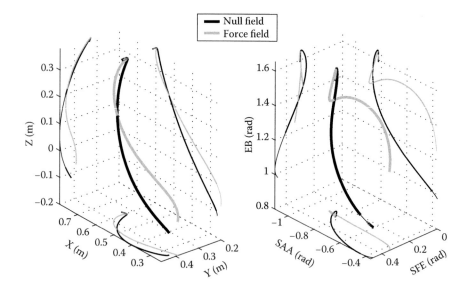

FIGURE 6.8 Simulated results. (Left) Hand of the rigid-body arm model, but without task-space compensation of the force field. A null field trial (black) is plotted with a force field trial (gray). (Right) Joint space trajectories (SAA, HR, and EB). Without compensating for the task-space forces produced by the force field disturbance, the controller cannot maintain the desired hand trajectory.

devices, a model-based controller with carefully tuned parameters accomplished a surprising quality of gravity, Coriolis, and inertia compensation, such that the robot did not alter the movement of a human inserted into the exoskeleton structure too much. We demonstrated the usefulness of this new experimental platform in a behavioral experiment, where human subjects were exposed to a novel dynamic environment in the form of joint-space force fields. The experiment provides insight into the planning and control of human reaching movements. The results suggest that humans do not employ invariant desired joint-space trajectories, but rather, subjects are primarily concerned with the trajectory of the hand. Therefore, they only compensate for task-relevant disturbances, as opposed to learning and compensating for the complete disturbance. We suggest that the nervous system may employ a form of operational space control. We show how an operational space controller applied to a simulated rigid-body arm model can produce similar motion to the observed results in human experiments.

6.3   WHOLE BODY KINEMATICS

The remainder of the chapter focuses on extending the modeling tools developed for task-space control to consider the whole body of humans and humanoids. A whole body human or humanoid robot complicates modeling and control. First, humanoid motion is inherently *underactuated*, having less control input than motion output. If a robot is not physically attached to the world, we need to use a *floating base* representation, such that the full configuration of the robot is described with respect to an inertial frame of reference. We attach a base reference frame to the robot, and assume that the base frame can move freely relative to the fixed inertial frame. Because the base frame cannot be moved with direct actuation, we have to plan our motion and control considering the limitations of the passive joints. Second, humanoids are typically *constrained*, for example, in environmental contact, such as with two feet on the floor. The constraints interject contact forces into the system, which must be compensated for when planning motion and resolving tasks. Constraints can be useful as well, as we show for resolving underactuation. In this and the next two sections, we demonstrate how to achieve the three control approaches shown in Figure 6.2 for whole-body humanoids in contact with the environment.

6.3.1   PRELIMINARIES: FLOATING BASE KINEMATICS

Again, we assume the robot has n degrees-of-freedom (DOFs) and is represented by the configuration vector $\mathbf{q} \in \mathbb{R}^n$. Some of the DOFs may be underactuated, in which case we have p active joints and $l = n - p$ passive joints. We always write the vector $\mathbf{q}$ assuming the first p components are active. For example, in the case of a humanoid robot with all actuated joints other than the floating base, we have $l = 6$, and $\mathbf{q}$ is decomposed as

$$\mathbf{q} = \begin{bmatrix} \mathbf{q}_a^T & \mathbf{x}_b^T \end{bmatrix}^T \tag{6.17}$$

where $\mathbf{q}_a \in \mathbb{R}^p$ is the joint configuration of the robot's actuated joints and $\mathbf{x}_b \in SE(3)$ is the position and orientation of the coordinate system attached to the robot's base

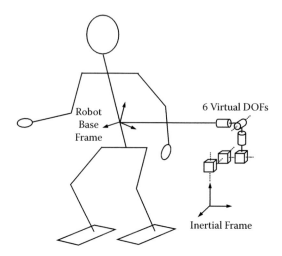

FIGURE 6.9 The base frame attached to the robot is connected to the inertial frame via six unactuated virtual DOFs.

frame, and measured with respect to the fixed inertial frame. Figure 6.9 illustrates this representation by showing the six virtual degrees-of-freedom attached from the inertial frame to the robot base frame.

Additionally, the robot may be in rigid contact with the environment, for example, a standing humanoid with two feet on the floor. To represent such contact, we write that the robot has k linearly independent constraints at positions and/or orientations in the vector $\mathbf{x}_C \in \mathbb{R}^k$. By definition, these constraints undergo zero acceleration with respect to the inertial frame, and we write this condition as $\ddot{\mathbf{x}}_C = \mathbf{J}_C \ddot{\mathbf{q}} + \dot{\mathbf{J}}_C \dot{\mathbf{q}} = \mathbf{0}$, where $\mathbf{J}_C$ is the constraint Jacobian. Note that constraints of this form can either be holonomic or nonholonomic [19].

6.3.2 FLOATING BASE INVERSE KINEMATICS

We assume we have the desired motion for the end-effector, written as $\mathbf{x}_d(t), \dot{\mathbf{x}}_d(t), \ddot{\mathbf{x}}_d(t)$. Recall, for a fully actuated fixed-base robot (such as a classical manipulator), we can compute a joint velocity vector to realize the desired task velocity by:

$$\dot{\mathbf{q}}_d = \mathbf{J}^+ \dot{\mathbf{x}}_d + \left(\mathbf{I} - \mathbf{J}^+ \mathbf{J}\right) \dot{\mathbf{q}}_0, \tag{6.18}$$

where $\mathbf{J}^+$ is the Moore–Penrose (MP) pseudoinverse of $\mathbf{J}$ and $\dot{\mathbf{q}}_0$ is an arbitrary velocity vector. If $\mathbf{J}$ is full row rank, we can compute the MP pseudoinverse by $\mathbf{J}^+ = \mathbf{J}^T \left(\mathbf{J}\mathbf{J}^T\right)^{-1}$. Similarly, we can solve for joint accelerations as

$$\ddot{\mathbf{q}}_d = \mathbf{J}^+ \left(\ddot{\mathbf{x}}_d - \dot{\mathbf{J}}\dot{\mathbf{q}}\right) + \left(\mathbf{I} - \mathbf{J}^+ \mathbf{J}\right) \ddot{\mathbf{q}}_0 \tag{6.19}$$

As we use the MP pseudo-inverse in the above equations, we compute the minimum vector norm for $\dot{\mathbf{q}}_d$ and $\ddot{\mathbf{q}}_d$ (assuming $\dot{\mathbf{q}}_0$ and $\ddot{\mathbf{q}}_0$ are zero), for example, the smallest possible velocity vector that will achieve $\dot{\mathbf{x}}_d$.* Of course, other pseudo-inverses could be used and would resolve redundancy in a different manner. A weighted pseudo-inverse, for example, can be used to redistribute the contribution of specific joints. An inertia weighted pseudo-inverse (see Section 6.5) would minimize the overall kinetic energy of the system.

Once $\dot{\mathbf{q}}_d$ or $\ddot{\mathbf{q}}_d$ are computed, we can obtain $\mathbf{q}_d$ by numerical integration, for example:

$$\mathbf{q}_d(t) = \mathbf{q}_d(t - \Delta t) + \Delta t \dot{\mathbf{q}}_d(t) \tag{6.20}$$

where Δt is the integration time step. Then as diagrammed in Figure 6.2A, the desired joint trajectory can be realized on the robot via a PD controller:

$$\tau = \mathbf{k}_D^T (\dot{\mathbf{q}}_d - \dot{\mathbf{q}}) + \mathbf{k}_P^T (\mathbf{q}_d - \mathbf{q}) \tag{6.21}$$

where $\mathbf{k}_P$ and $\mathbf{k}_D$ are vectors of position and velocity gains, respectively, and τ is the vector of joint torque commands.

The problem, however, for systems with a floating base or other underactuation, is insufficient control authority to realize arbitrary joint trajectories via PD control. Equation (6.21) computes torque commands for passive joints. Instead, if we can constrain the system, using environmental contact, we can potentially reduce the dimensionality of the system to a subspace where we have full control authority. In a sense, the underactuated system becomes fully actuated within this subspace. In general, if the system has l passive joints, the system requires $k \geq l$ constraints to have full controllability within the reduced $n - k$ dimensional space. If such constraints exist, we can compute desired velocities for only the active joints $\mathbf{q}_a$, by first augmenting the task Jacobian with the constraint Jacobian $\mathbf{J}_C$, and computing:

$$\dot{\mathbf{q}}_a = \begin{bmatrix} \mathbf{I}_{p \times p} & \mathbf{0}_{p \times l} \end{bmatrix} \begin{bmatrix} \mathbf{J}_C \\ \mathbf{J} \end{bmatrix}^+ \begin{bmatrix} \mathbf{0}_{k \times 1} \\ \dot{\mathbf{x}}_d \end{bmatrix} \tag{6.22}$$

These velocities can then be achieved, by PD control, for example. The desired end-effector velocity ($\dot{\mathbf{x}}_d$) will be realized, provided the following sufficient conditions are satisfied:

1. $\ddot{\mathbf{x}}_C = \mathbf{0}$ is maintained by the contacts at the constraint locations. This implies that there is sufficient friction and rigidity at the constraint locations to prevent acceleration relative to the inertial frame.[†]

---

* Here for simplicity we assume that the $\dot{\mathbf{q}}$ and $\ddot{\mathbf{q}}$ are homogeneous (e.g., all components are represented in the same units, e.g., rad/s for joint velocity or rad/s^2 for joint acceleration). This implies that a suitable metric can be defined and a minimum norm exists. If the units of these vectors are heterogeneous, for example, a mixture of joint and linear velocities, a weighted pseudo-inverse needs to be used. Full details can be found in [19]

[†] Note that we are assuming the constraint is *multilateral*, whereas a typical floor or wall contact will be *unilateral* (acting in only one direction). For now, we assume that the desired task trajectories and the controller are designed to maintain unilateral contacts, and we do not explicitly consider the unilateral constraint case.

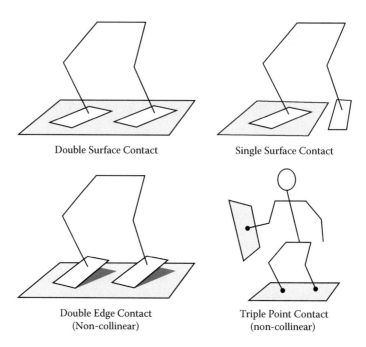

FIGURE 6.10 Examples of systems that are adequately constrained, and remain fully actuated within a dimensionally reduced space: Rank $\left(\mathbf{J}_C^{\partial \mathbf{x}_b} \right) = l$.

2. The system is not overconstrained. The augmented matrix $\left[\mathbf{J}_C^T \ \mathbf{J}^T \right]^T$ should remain full row rank. If the matrix is not full row rank, there are two possiblities: (1) there exist tasks or constraints that are linearly dependent on one another. In which case, we can reformulate $\mathbf{x}$ and $\mathbf{x}_C$ such that all tasks are linearly independent and the matrix becomes full row rank. Alternatively, (2) there do not exist enough degrees-of-freedom in the system to realize all desired tasks and constraints. Note that this condition necessarily implies that $m + k \le n$, and task choice will be limited by the number of constraints.

3. The system is not underconstrained. Recalling the notation introduced in (6.17): $\mathbf{J}_C = [\, \mathbf{J}_C^{\partial \mathbf{q}_a} \ \mathbf{J}_C^{\partial \mathbf{x}_b} \,]$, the constraint Jacobian related to the unactuated DOFs, $\mathbf{J}_C^{\partial \mathbf{x}_b}$, must have a rank equal to l (exactly the number of passive DOFs). If the rank is less than l, the system will be underactuated. Figure 6.10 shows examples where $\text{Rank}(\mathbf{J}_C^{\partial \mathbf{x}_b}) = l$, and the system remains fully actuated within the reduced dimensional space. Figure 6.11 has examples where $\text{Rank}(\mathbf{J}_C^{\partial \mathbf{x}_b}) < l$, and the system is underactuated. Methods for controlling underactuated systems are presented in Section 6.5.2.

Next we prove that an actuated joint velocity $\dot{\mathbf{q}}_a$ computed by (6.22) will realize the desired end-effector velocity $\dot{\mathbf{x}}_d$, assuming the above three conditions are satisfied.

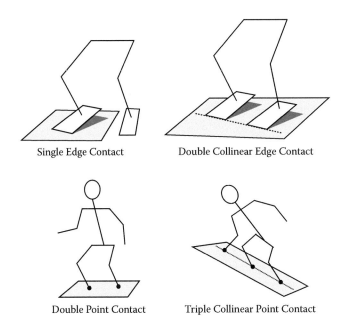

Single Edge Contact   Double Collinear Edge Contact

Double Point Contact   Triple Collinear Point Contact

FIGURE 6.11 Examples of systems that are underactuated: Rank $\left(\mathbf{J}_C^{\partial \mathbf{x}_b} \right) < l$.

Proof. If $\left[\, \mathbf{J}_C^T \, \mathbf{J}^T \, \right]^T$ is full row rank (condition 2), its right pseudo-inverse exists, and we can compute $\dot{\mathbf{q}}$ as

$$\dot{\mathbf{q}} = \begin{bmatrix} \mathbf{J}_C \\ \mathbf{J} \end{bmatrix}^T \left(\begin{bmatrix} \mathbf{J}_C \\ \mathbf{J} \end{bmatrix} \begin{bmatrix} \mathbf{J}_C \\ \mathbf{J} \end{bmatrix}^T \right)^{-1} \begin{bmatrix} \mathbf{0}_{k \times 1} \\ \dot{\mathbf{x}}_d \end{bmatrix} \tag{6.23}$$

Assuming that this $\dot{\mathbf{q}}$ can be fully realized by some controller, and contact points do not slip (condition 1), it is trivial to verify:

$$\begin{bmatrix} \dot{\mathbf{x}}_C \\ \dot{\mathbf{x}} \end{bmatrix} = \begin{bmatrix} \mathbf{J}_C \\ \mathbf{J} \end{bmatrix} \begin{bmatrix} \mathbf{J}_C \\ \mathbf{J} \end{bmatrix}^T \left(\begin{bmatrix} \mathbf{J}_C \\ \mathbf{J} \end{bmatrix} \begin{bmatrix} \mathbf{J}_C \\ \mathbf{J} \end{bmatrix}^T \right)^{-1} \begin{bmatrix} \mathbf{0}_{k \times 1} \\ \dot{\mathbf{x}}_d \end{bmatrix} = \begin{bmatrix} \mathbf{0}_{k \times 1} \\ \dot{\mathbf{x}}_d \end{bmatrix} \tag{6.24}$$

resulting in $\dot{\mathbf{x}} = \dot{\mathbf{x}}_d$ and $\dot{\mathbf{x}}_C = \mathbf{0}$ (the desired task velocity is achieved while constraints are maintained). The next part of the proof verifies that $\dot{\mathbf{q}}$ can be realized by only explicitly controlling $\dot{\mathbf{q}}_a$. The proof relies on the fact that there exists a unique $\dot{\mathbf{q}}$ for a given $\dot{\mathbf{q}}_a$. From (6.24) we know $\dot{\mathbf{q}}$ will be in the null space of the constraints, so we can write:

$$\mathbf{J}_C \dot{\mathbf{q}} = \mathbf{J}_C^{\partial \mathbf{q}_a} \dot{\mathbf{q}}_a + \mathbf{J}_C^{\partial \mathbf{x}_b} \dot{\mathbf{x}}_b = \mathbf{0} \tag{6.25}$$

and solve for $\dot{\mathbf{x}}_b$:

$$\dot{\mathbf{x}}_b = - \left[\mathbf{J}_C^{\partial \mathbf{x}_b} \right]^{+} \mathbf{J}_C^{\partial \mathbf{q}_a} \dot{\mathbf{q}}_a \tag{6.26}$$

If and only if (6.26) has a unique solution, will $\dot{\mathbf{q}}$ be unique for any given $\dot{\mathbf{q}}_a$. Note that $\mathbf{J}_C^{\partial \mathbf{x}_b}$ is of size $k \times l$. If $k < l$, Equation (6.26) has potentially an infinite number of solutions, and $\dot{\mathbf{q}}$ cannot be unique. However, if $k \geq l$ and $\mathbf{J}_C^{\partial \mathbf{x}_b}$ is full column rank, the rank is l, and a unique solution is guaranteed [29].* The unique solution is given by the left pseudo-inverse $(A^+ = (A^T A)^{-1} A^T)$. Thus, when $\text{Rank}\left(\mathbf{J}_C^{\partial \mathbf{x}_b}\right) = l$, we can represent $\dot{\mathbf{q}}$ unambiguously as a function of $\dot{\mathbf{q}}_a$:

$$\dot{\mathbf{q}} = \left[\mathbf{I}_{n \times n} - \left[\mathbf{J}_C^{\partial \mathbf{x}_b}{}^T \mathbf{J}_C^{\partial \mathbf{x}_b} \right]^{-1} \mathbf{J}_C^{\partial \mathbf{x}_b}{}^T \mathbf{J}_C^{\partial \mathbf{q}_a} \right] \dot{\mathbf{q}}_a \tag{6.27}$$

and although we compute a full $\dot{\mathbf{q}}$ using (6.23), we only need to control for $\dot{\mathbf{q}}_a$. ∎

Similarly to the velocity case, we can also compute the actuated joint accelerations for achieving $\ddot{\mathbf{x}}_d$:

$$\ddot{\mathbf{q}}_a = \begin{bmatrix} \mathbf{I}_{p \times p} & \mathbf{0}_{p \times l} \end{bmatrix} \begin{bmatrix} \mathbf{J}_C \\ \mathbf{J} \end{bmatrix}^+ \left(\begin{bmatrix} \mathbf{0}_{k \times 1} \\ \ddot{\mathbf{x}}_d \end{bmatrix} - \begin{bmatrix} \dot{\mathbf{J}}_C \\ \dot{\mathbf{J}} \end{bmatrix} \dot{\mathbf{q}} \right) \tag{6.28}$$

and assuming the same three conditions above are satisfied, $\ddot{\mathbf{x}} = \ddot{\mathbf{x}}_d$ will be realized. The proof follows the same logic as the velocity case.

6.3.3 HUMANOID MULTITASK CONTROL USING INVERSE KINEMATICS

In this section, we relate floating base inverse kinematics specifically to humanoid control. The high number of DOFs of humanoids allows for a variety of tasks to be completed simultaneously, for example, locomotion while manipulating a tool and orienting the head to look at an object. However, in order to maintain stability and controllability of the robot, it is critical to manage the different tasks and constraints in an appropriate hierarchical manner. Therefore, as similarly done in Reference [23], we categorize the robot's desired tasks into one of three prioritization groups:

1. *Critical tasks*: Essential tasks that must succeed. All other tasks depend on the success of these tasks.
2. *Functional tasks*: Tasks we want to accomplish with a high level of accuracy, but their success or failure generally does not affect other tasks. This group could potentially have additional sublevels of hierarchy within it.
3. *Extraneous tasks*: These tasks are used for resolving any remaining redundancy (such that the robot does not drift in null space). Success of these tasks is generally irrelevant.

Once categorized in this manner, the tasks will be controlled via a task prioritization framework.

As emphasized in the previous section, in order to be able to use (6.22) or (6.28) to control arbitrary tasks, it is critical that constraints are maintained at all times. If

---

* There is still a possibility that no solution exists. However, we are also assuming (6.25) is true, so we know at least one solution must exist.

a contact point ever has a nonzero acceleration, then the robot's base will undergo motion independent of the joints, and as a consequence we lose the controllability of all tasks. Thus, when deriving a desired acceleration for the robot, a critical task is to maintain zero acceleration at all contact points. Secondly, balance is obviously critical, inasmuch as a fall can be disastrous. Thus we always place emphasis on static stability (maintaining the COG over the support polygon). For the purpose of this section, we assume that the robot is in double surface contact with the environment (both feet are flat on the floor), and we define our critical task Jacobian as

$$
\mathbf{J}_A = \begin{bmatrix} \mathbf{J}_{\text{rightfoot}} \\ \mathbf{J}_{\text{leftfoot}} \\ \mathbf{J}_{\text{cog},xy} \end{bmatrix}
\tag{6.29}
$$

where $\mathbf{J}_{\text{rightfoot}}$ and $\mathbf{J}_{\text{leftfoot}}$ are the Jacobians for the position and orientation of the two feet, and $\mathbf{J}_{\text{cog},xy}$ is the Jacobian for the projection of the COG onto the XY plane (assuming the gravity vector is in the Z direction). Because we plan to compute a pseudo-inverse during real-time control, as with (6.22), it is critical for robot safety that we avoid any singularities of $\mathbf{J}_A$. Singularities will occur when two or more tasks are in conflict, that is, when the task of controlling the XY position of the COG conflicts with the task of maintaining the position of either foot. This particular situation would only occur when the robot is fully extended across the floor (i.e., the robot has already fallen), or has its legs in a full split position. Thus in practice, as long as the robot is upright, we have no real concerns of $\mathbf{J}_A$ reaching a singularity. Note that fully extended knees are not a singular configuration in this case, and balance can still be maintained (provided the ankle joints have sufficient torque).

For a functional task we may want to control hand motion, for example, and thus we define the functional task Jacobian as

$$
\mathbf{J}_B = \begin{bmatrix} \mathbf{J}_{\text{righthand}} \end{bmatrix}
\tag{6.30}
$$

where $\mathbf{J}_{\text{righthand}}$ is the Jacobian for the right hand.

Finally we define our lowest priority (extraneous) tasks as

$$
\mathbf{J}_E = \begin{bmatrix} \mathbf{J}_{\text{rpy}} \\ \mathbf{J}_{\text{cog},z} \end{bmatrix}
\tag{6.31}
$$

where $\mathbf{J}_{\text{rpy}}$ dictates base orientation (in roll, pitch, yaw angles), and $\mathbf{J}_{\text{cog},z}$ for the height of the COG above the ground. Note that it is necessary to control these variables in order to resolve any leftover redundancy, and prevent the robot from drifting in null space.

Next we define the vectors of desired end-effector velocities, noting that $\dot{\mathbf{x}}_{\text{rightfoot}} = \dot{\mathbf{x}}_{\text{leftfoot}} = \mathbf{0}$ in order to maintain constraints:

$$
\dot{\mathbf{x}}_A = \begin{bmatrix} \mathbf{0} \\ \mathbf{0} \\ \dot{\mathbf{x}}_{\text{cog},xy} \end{bmatrix} \quad \dot{\mathbf{x}}_B = \begin{bmatrix} \dot{\mathbf{x}}_{\text{righthand}} \end{bmatrix}, \quad \dot{\mathbf{x}}_E = \begin{bmatrix} \dot{\mathbf{x}}_{\text{rpy}} \\ \dot{\mathbf{x}}_{\text{cog},z} \end{bmatrix}
\tag{6.32}
$$

Finally, we compute the desired joint velocities as

$$\dot{\mathbf{q}}_a = \begin{bmatrix} \mathbf{I}_{(n-6)\times(n-6)} \\ \mathbf{0}_{6\times(n-6)} \end{bmatrix}^T \left(\mathbf{J}_A^+ \dot{\mathbf{x}}_A + \mathbf{N}_A \mathbf{J}_B^+ \dot{\mathbf{x}}_B + \mathbf{N}_{AB} \mathbf{J}_E^+ \dot{\mathbf{x}}_E \right) \qquad (6.33)$$

where $\mathbf{N}_A = \mathbf{I} - \mathbf{J}_A^+ \mathbf{J}_A$ is the null space projection matrix of $\mathbf{J}_A$ and

$$\mathbf{N}_{AB} = \mathbf{I} - \begin{bmatrix} \mathbf{J}_A \\ \mathbf{J}_B \end{bmatrix}^+ \begin{bmatrix} \mathbf{J}_A \\ \mathbf{J}_B \end{bmatrix} \qquad (6.34)$$

projects into the intersection of the $\mathbf{J}_A$ and $\mathbf{J}_B$ null spaces. Note that we conduct task prioritization via null-space projections, which as a consequence does not track secondary tasks as accurately as other task prioritization frameworks [3]. However (6.33) uses the only task prioritization framework that will guarantee the success of $\dot{\mathbf{x}}_A$ (our critical tasks) for all time (except at rare singularities of $\mathbf{J}_A$, as discussed previously) and does not suffer from the problem of algorithmic singularities as the other approaches do [17,26], or require a less accurate damped pseudo-inverse [5]. For more discussion of these issues in relation to floating-base kinematics, please refer to Reference [15].*

Extending (6.33) to the second-order acceleration case, we have:

$$\ddot{\mathbf{q}} = \mathbf{J}_A^+ \left(\ddot{\mathbf{x}}_A - \dot{\mathbf{J}}_A \dot{\mathbf{q}} \right) + \mathbf{N}_A \mathbf{J}_B^+ \left(\ddot{\mathbf{x}}_B - \dot{\mathbf{J}}_B \dot{\mathbf{q}} \right) + \mathbf{N}_{AB} \mathbf{J}_E^+ \left(\ddot{\mathbf{x}}_E - \dot{\mathbf{J}}_E \dot{\mathbf{q}} \right) \qquad (6.35)$$

and $\ddot{\mathbf{q}}_a = \begin{bmatrix} \mathbf{I}_{(n-6)\times(n-6)} & \mathbf{0}_{(n-6)\times 6} \end{bmatrix} \ddot{\mathbf{q}}$. Note that computing (6.35) requires knowledge of $\dot{\mathbf{q}}$. It can be obtained by measurement from noisy sensors or by numerical integration. Either way, the result may not be consistent with the constraints, which in turn will make $\ddot{\mathbf{q}}$ constraint inconsistent. In order to enforce constraint consistency, the $\dot{\mathbf{q}}$ to be used in (6.35) should be first filtered through the null space of the constraints $(\mathbf{I} - \mathbf{J}_C^+ \mathbf{J}_C)$.

After computing $\dot{\mathbf{q}}_a$ from (6.33), $\mathbf{q}_a$ can be computed via numerical integration:

$$\mathbf{q}_a(t) = \mathbf{q}_a(t - \Delta t) + \Delta t \dot{\mathbf{q}}_a(t) \qquad (6.36)$$

where t is the current time step, and Δt is the control cycle period. Similarly, if Equation (6.35) is used:

$$\dot{\mathbf{q}}_a(t) = \dot{\mathbf{q}}_a(t - \Delta t) + \Delta t \ddot{\mathbf{q}}_a(t) \qquad (6.37)$$

$$\mathbf{q}_a(t) = \mathbf{q}_a(t - \Delta t) + \Delta t \dot{\mathbf{q}}_a(t) \qquad (6.38)$$

Subsequently, a PD controller can be used to realize the desired joint trajectories. Alternatively, we can compute control input directly from joint accelerations, via inverse dynamics control, and that is the topic of the next section.

---

* Also, because technically $\mathbf{J}_A$ and $\mathbf{J}_B$ could conflict, we should replace $\mathbf{N}_{AB}$ with $\mathbf{N}_A \mathbf{N}_{AB}^\lambda$ where $\mathbf{N}_{AB}^\lambda$ uses a damping for the pseudo-inverse; see Reference [15] for more detail.

6.4 WHOLE BODY INVERSE DYNAMICS

The whole body rigid-body inverse-dynamics equation is written as

$$\mathbf{M}(\mathbf{q})\ddot{\mathbf{q}} + \mathbf{h}(\mathbf{q}, \dot{\mathbf{q}}) = \mathbf{B}\tau + \mathbf{J}_C^{T}(\mathbf{q})\lambda \tag{6.39}$$

where $\mathbf{M}(\mathbf{q}) \in \mathbb{R}^{n \times n}$ is the inertia matrix; $\mathbf{h}(\mathbf{q}, \dot{\mathbf{q}}) \in \mathbb{R}^n$ is the vector of centripetal, Coriolis, and gravity forces; $\tau \in \mathbb{R}^n$ is the vector of joint torques; $\lambda \in \mathbb{R}^k$ is the vector of k linearly independent constraint forces; and

$$\mathbf{B} = \begin{bmatrix} \mathbf{I}_p & \mathbf{0} \\ \mathbf{0} & \mathbf{0} \end{bmatrix} \tag{6.40}$$

is the projector into actuated joint space ($\mathbf{I}_p$ is the p-dimensional identity matrix). We assume the first p rows of (6.39) represent the actuated DOFs. Our goal is to compute the joint torques that will realize some desired joint motion $\ddot{\mathbf{q}}_d$, for example, as diagrammed in Figure 6.2B.

Recall that for a fully actuated fixed-base robot (such as a classical manipulator), we will have $\mathbf{B} = \mathbf{I}$ and $\lambda = \mathbf{0}$, and we can solve for joint torques to realize a given joint acceleration $\ddot{\mathbf{q}}_d$ by:

$$\tau = \mathbf{M}\ddot{\mathbf{q}}_d + \mathbf{h} \tag{6.41}$$

and by plugging (6.41) into (6.39) (again with $\mathbf{B} = \mathbf{I}$, $\lambda = \mathbf{0}$) it is easy to verify $\ddot{\mathbf{q}} = \ddot{\mathbf{q}}_d$. Once again, however, underactuation will be problematic, and prevent us from doing so. Additionally, contact forces (which are assumed to be unknown) need to be compensated for as well.

As outlined by Reference [1], we can project (6.39) into the null space of the constraints, eliminating the constraint forces from the dynamics equation, and write the equation of projected inverse dynamics:

$$\mathbf{PM}\ddot{\mathbf{q}} + \mathbf{Ph} = \mathbf{PB}\tau \tag{6.42}$$

where $\mathbf{P}$ is an orthogonal projection operator, such that $\mathbf{PJ}_C^T = \mathbf{0}$ and $\mathbf{P} = \mathbf{P}^2 = \mathbf{P}^T$. Also, $\mathbf{P}$ is readily computable from the constraint Jacobian: $\mathbf{P} = \mathbf{I} - \mathbf{J}_C^+\mathbf{J}_C$ (where $+$ indicates the Moore–Penrose pseudo-inverse). Note that $\mathbf{P}$ depends only on kinematic parameters and unlike other approaches that attempt to maintain dynamic consistency with constraints, it does not require the inertia matrix. Then for a $\ddot{\mathbf{q}}_d$ that respects the constraints (e.g., as computed by (6.28)), we can achieve $\ddot{\mathbf{q}} = \ddot{\mathbf{q}}_d$ using the equation:

$$\tau = [\mathbf{PB}]^+ (\mathbf{M}\ddot{\mathbf{q}}_d + \mathbf{h}) \tag{6.43}$$

To show that this is true, we use an approach based on the QR decomposition of $\mathbf{J}_C$.

6.4.1 ORTHOGONAL DECOMPOSITION OF THE CONSTRAINT JACOBIAN

We start with the rigid-body dynamics equation (6.44) written as

$$\mathbf{M}(\mathbf{q})\ddot{\mathbf{q}} + \mathbf{h}(\mathbf{q}, \dot{\mathbf{q}}) = \mathbf{S}^T \tau_a + \mathbf{J}_C^T(\mathbf{q})\lambda \tag{6.44}$$

where $\tau_a = \mathbf{S}\tau$ is the vector of actuated joint torques and $\mathbf{S} = \begin{bmatrix} \mathbf{I}_{p \times p} & \mathbf{0}_{p \times l} \end{bmatrix}$.

If we assume Rank $(\mathbf{J}_C) = k$ (i.e., there are k linearly independent constraints and $\mathbf{J}_C$ is full row rank), we can then compute the QR decomposition of $\mathbf{J}_C^T$:

$$\mathbf{J}_C^T = \mathbf{Q} \begin{bmatrix} \mathbf{R} \\ \mathbf{0} \end{bmatrix} \tag{6.45}$$

where $\mathbf{Q}$ is orthogonal ($\mathbf{Q}\mathbf{Q}^T = \mathbf{Q}^T\mathbf{Q} = \mathbf{I}$), and $\mathbf{R}$ is an upper triangle matrix of rank k. Additionally, if $\mathbf{R}$ is restricted to have all positive diagonal elements, then $\mathbf{Q}$ and $\mathbf{R}$ are unique. Multiplying (6.39) by $\mathbf{Q}^T$ allows us to decompose the rigid body dynamics into two equations:

$$\mathbf{S}_c\mathbf{Q}^T (\mathbf{M}\ddot{\mathbf{q}} + \mathbf{h}) = \mathbf{S}_c\mathbf{Q}^T\mathbf{S}^T \tau_a + \mathbf{R}\lambda \tag{6.46}$$

$$\mathbf{S}_u\mathbf{Q}^T (\mathbf{M}\ddot{\mathbf{q}} + \mathbf{h}) = \mathbf{S}_u\mathbf{Q}^T\mathbf{S}^T \tau_a \tag{6.47}$$

where

$$\mathbf{S}_c = \begin{bmatrix} \mathbf{I}_{k \times k} & \mathbf{0}_{k \times (n-k)} \end{bmatrix} \tag{6.48}$$

$$\mathbf{S}_u = \begin{bmatrix} \mathbf{0}_{(n-k) \times k} & \mathbf{I}_{(n-k) \times (n-k)} \end{bmatrix} \tag{6.49}$$

are used to select the upper and lower portions of the full equation. In the proof below we show that if the motion remains within the null space of the constraints (i.e., as computed by (6.28), then Equations (6.46) and (6.47) will be independent. Both equations will represent the full dynamics of the robot. Because (6.47) contains no dependence on constraint forces, it is simple to compute control torques using (6.47) and a pseudo-inverse:

$$\tau_a = \left(\mathbf{S}_u\mathbf{Q}^T\mathbf{S}^T\right)^+ \mathbf{S}_u\mathbf{Q}^T [\mathbf{M}\ddot{\mathbf{q}}_d + \mathbf{h}] . \tag{6.50}$$

Additionally, contact forces can be computed using the upper equation (6.46):

$$\lambda = \mathbf{R}^{-1}\mathbf{S}_c\mathbf{Q}^T \left[\mathbf{M}\ddot{\mathbf{q}}_d + \mathbf{h} - \mathbf{S}^T \tau_a\right] \tag{6.51}$$

Note that the modeling errorprone inertia matrix is used only once, in a noninverted form, and the time derivative of the constraint Jacobian ($\dot{\mathbf{J}}_C$) is not required.

Proof. Here we prove the independence of (6.46) and (6.47), assuming $\mathbf{J}_C\dot{\mathbf{q}} = \mathbf{0}$. We begin by multiplying (6.44) by $\mathbf{Q}^T$:

$$\mathbf{Q}^T\mathbf{M}\ddot{\mathbf{q}} + \mathbf{Q}^T\mathbf{h} = \mathbf{Q}^T\mathbf{S}^T \tau_a + \begin{bmatrix} \mathbf{R} \\ \mathbf{0} \end{bmatrix} \lambda \tag{6.52}$$

and define a new coordinate system $\mathbf{p}$ such that:

$$\dot{\mathbf{q}} = \mathbf{Q}\dot{\mathbf{p}}, \quad \ddot{\mathbf{q}} = \mathbf{Q}\ddot{\mathbf{p}} + \dot{\mathbf{Q}}\dot{\mathbf{p}} \tag{6.53}$$

We then rewrite (6.52) as

$$\mathbf{Q}^T\mathbf{M}\mathbf{Q}\ddot{\mathbf{p}} + \mathbf{Q}^T\mathbf{M}\dot{\mathbf{Q}}\dot{\mathbf{p}} + \mathbf{Q}^T\mathbf{h} = \mathbf{Q}^T\mathbf{S}^T \tau_a + \begin{bmatrix} \mathbf{R} \\ \mathbf{0} \end{bmatrix} \lambda \tag{6.54}$$

Next, we decompose $\mathbf{p}$ as follows,

$$\mathbf{p} = \begin{bmatrix} \mathbf{p}_c \\ \mathbf{p}_u \end{bmatrix} \tag{6.55}$$

where $\mathbf{p}_c \in \mathbb{R}^k$ and $\mathbf{p}_u \in \mathbb{R}^{n-k}$. If $\dot{\mathbf{q}}$ is in the null space of $\mathbf{J}_C$:

$$\mathbf{J}_C \dot{\mathbf{q}} = \begin{bmatrix} \mathbf{R}^T & \mathbf{0} \end{bmatrix} \mathbf{Q}^T \dot{\mathbf{q}} = \begin{bmatrix} \mathbf{R}^T & \mathbf{0} \end{bmatrix} \begin{bmatrix} \dot{\mathbf{p}}_c \\ \dot{\mathbf{p}}_u \end{bmatrix} = \mathbf{0} \tag{6.56}$$

implying that $\dot{\mathbf{p}}_c = \mathbf{0}$. If we differentiate (6.56):

$$\begin{bmatrix} \mathbf{R}^T & \mathbf{0} \end{bmatrix} \begin{bmatrix} \ddot{\mathbf{p}}_c \\ \ddot{\mathbf{p}}_u \end{bmatrix} + \begin{bmatrix} \dot{\mathbf{R}}^T & \mathbf{0} \end{bmatrix} \begin{bmatrix} \dot{\mathbf{p}}_c \\ \dot{\mathbf{p}}_u \end{bmatrix} = \mathbf{0} \tag{6.57}$$

Because $\dot{\mathbf{p}}_c = \mathbf{0}$, we must also have $\ddot{\mathbf{p}}_c = \mathbf{0}$. Finally we can conclude that:

$$\dot{\mathbf{p}}_c = \ddot{\mathbf{p}}_c = \mathbf{0}, \quad \mathbf{p}_c = \mathbf{k} \tag{6.58}$$

where $\mathbf{k}$ is some constant vector (it is unknown, but irrelevant). Thus we conclude that $\mathbf{p}_c$ plays no role in the system dynamics. This effectively decouples (6.54) into two independent equations, which we write as

$$\mathbf{S}_c \left(\mathbf{Q}^T \mathbf{M} \mathbf{Q} \ddot{\mathbf{p}} + \mathbf{Q}^T \mathbf{M} \dot{\mathbf{Q}} \dot{\mathbf{p}} + \mathbf{Q}^T \mathbf{h} \right) = \mathbf{S}_c \mathbf{Q}^T \mathbf{S}^T \tau_a + \mathbf{R} \lambda \tag{6.59}$$

$$\mathbf{S}_u \left(\mathbf{Q}^T \mathbf{M} \mathbf{Q} \ddot{\mathbf{p}} + \mathbf{Q}^T \mathbf{M} \dot{\mathbf{Q}} \dot{\mathbf{p}} + \mathbf{Q}^T \mathbf{h} \right) = \mathbf{S}_u \mathbf{Q}^T \mathbf{S}^T \tau_a \tag{6.60}$$

Finally, transforming back into our original coordinate system, we substitute $\dot{\mathbf{p}} = \mathbf{Q}^T \dot{\mathbf{q}}$ and $\ddot{\mathbf{p}} = \mathbf{Q}^T \ddot{\mathbf{q}} - \mathbf{Q}^T \dot{\mathbf{Q}} \mathbf{Q}^T \dot{\mathbf{q}}$ into the above and write the following,

$$\mathbf{S}_c \mathbf{Q}^T \left(\mathbf{M} \ddot{\mathbf{q}} + \mathbf{h} \right) = \mathbf{S}_c \mathbf{Q}^T \mathbf{S}^T \tau_a + \mathbf{R} \lambda \tag{6.61}$$

$$\mathbf{S}_u \mathbf{Q}^T \left(\mathbf{M} \ddot{\mathbf{q}} + \mathbf{h} \right) = \mathbf{S}_u \mathbf{Q}^T \mathbf{S}^T \tau_a \tag{6.62}$$

∎

The proof has verified that the two equations (6.61) and (6.62) are dynamically independent of each other. This means we can represent the full dynamics of the system using either equation alone. Again, their independence relies on the assumption that our joint motion remains consistent with the constraints (i.e., $\ddot{\mathbf{q}}$ lives within the null space of the constraint Jacobian). We can ensure that we compute a desired joint acceleration $\ddot{\mathbf{q}}_d$ that is constraint consistent by using (6.28). Then we will be able to compute the torque commands required to realize $\ddot{\mathbf{q}}_d$ by using (6.62) and a pseudo-inverse:

$$\tau_a = \left(\mathbf{S}_u \mathbf{Q}^T \mathbf{S}^T \right)^+ \mathbf{S}_u \mathbf{Q}^T \left[\mathbf{M} \ddot{\mathbf{q}}_d + \mathbf{h} \right] \tag{6.63}$$

and the equation does not require any knowledge of the contact forces. Although this method produces the control input that will realize the task, it still requires an explicit inverse kinematics step. Alternatively, we can compute the task-relevant forces directly from the desired task. For this, we use operational space control.

6.5 WHOLE BODY OPERATIONAL-SPACE CONTROL

We have described how operational-space control is a method for decoupling task and task-irrelevant dynamics. Given a desired task, we can directly compute the control forces required to realize the task (without resorting to an inverse kinematics procedure). Additionally, we can compute the task-irrelevant forces that have no effect on the task. These null-space forces can be used to achieve secondary tasks or resolve redundancy. We have also shown how operational-space control can be a useful tool for modeling human movement. In our experiment, we showed how in a reaching movement, the subject only compensates for the task-relevant disturbances, and task-irrelevant disturbances are allowed to be pushed into the null space. Our challenge now is to extend this modeling technique into whole-body motion, where under-actuation and environmental contact are concerns. Following our work in [16] we derive operational-space control for constrained systems, and subsequently suggest how underactuated dynamics can be compensated for using null-space forces.

6.5.1 OPERATIONAL-SPACE DYNAMICS FOR CONSTRAINED SYSTEMS

To apply the operational-space control for humanoid robots, we once again have to cope with underactuation and constraints. First we derive operational-space dynamics for fully actuated but constrained systems, for example, a robot manipulator in contact with the environment. Underactuation is addressed in a subsequent section. We start with (6.42), the equation of projected inverse dynamics, and assume full actuation ($\mathbf{B} = \mathbf{I}$):

$$\mathbf{PM\ddot{q}} + \mathbf{Ph} = \mathbf{P\tau} \tag{6.64}$$

We wish to invert $\mathbf{PM}$ to solve for $\ddot{\mathbf{q}}$; however, because $\mathbf{P}$ is generally rank deficient, this term is not invertible. However, because our system is constrained, we have the following additional equations,

$$(\mathbf{I} - \mathbf{P})\dot{\mathbf{q}} = \mathbf{0} \tag{6.65}$$

$$(\mathbf{I} - \mathbf{P})\ddot{\mathbf{q}} = \mathbf{C\dot{q}} \tag{6.66}$$

where $\mathbf{C}$ is $(d/(dt))\mathbf{P}$ (e.g., $\mathbf{C} = -\mathbf{J}_C^+\dot{\mathbf{J}}_C$). Employing a trick used by Aghili, recognizing that (6.66) is orthogonal to (6.64), we can add the two equations:

$$\mathbf{M}_c\ddot{\mathbf{q}} + \mathbf{Ph} - \mathbf{C\dot{q}} = \mathbf{P\tau} \tag{6.67}$$

defining $\mathbf{M}_c = \mathbf{PM} + \mathbf{I} - \mathbf{P}$. Note, as discussed by Aghili, the choice of $\mathbf{M}_c$ for a given $\mathbf{q}$ is not unique, but is always invertible (provided $\mathbf{M}$ is invertible).* Next, we multiply (6.67) by $\mathbf{JM}_c^{-1}$, and replace $\mathbf{J\ddot{q}}$ with $\ddot{\mathbf{x}} - \dot{\mathbf{J}}\dot{\mathbf{q}}$:

$$\ddot{\mathbf{x}} - \dot{\mathbf{J}}\dot{\mathbf{q}} + \mathbf{JM}_c^{-1}(\mathbf{Ph} - \mathbf{C\dot{q}}) = \mathbf{JM}_c^{-1}\mathbf{P\tau} \tag{6.68}$$

---

* Other choices of $\mathbf{M}_c$ include $(\mathbf{M} + \mathbf{PM} - \mathbf{MP})$, $(\mathbf{PMP} + (\mathbf{I} - \mathbf{P})\mathbf{M}(\mathbf{I} - \mathbf{P}))$, and $(\mathbf{PM} + \alpha(\mathbf{I} - \mathbf{P}))$ for arbitrary scalar α. The definition of $\mathbf{C}$ will change with choice of $\mathbf{M}_c$.

A force at the end-effector $\mathbf{F}$ is mapped into joint torques via $\tau = \mathbf{J}^T \mathbf{F}$, and therefore we derive the constrained operational-space dynamics as

$$\mathbf{\Lambda}_c \ddot{\mathbf{x}} + \mathbf{\Lambda}_c \left(\mathbf{JM}_c^{-1} \mathbf{Ph} - \left(\dot{\mathbf{J}} + \mathbf{JM}_c^{-1} \mathbf{C} \right) \dot{\mathbf{q}} \right) = \mathbf{F} \tag{6.69}$$

where $\mathbf{\Lambda}_c = \left(\mathbf{JM}_c^{-1} \mathbf{PJ}^T \right)^{-1}$. Although $\mathbf{M}_c$ is not unique, $\mathbf{M}_c^{-1} \mathbf{P}$ will be unique (for a given $\mathbf{q}$),* and therefore (6.69) is also unique.

The operational space control equation for constrained (fully actuated) systems also takes the form of (6.7). However, we use $\mathbf{F}$ defined by (6.69) (again replacing $\ddot{\mathbf{x}}$ with our desired task acceleration $\ddot{\mathbf{x}}_d$), and using the following generalized inverse of $\mathbf{J}^T$,

$$\mathbf{J}^{T\#} = \left(\mathbf{JM}_c^{-1} \mathbf{PJ}^T \right)^{-1} \mathbf{JM}_c^{-1} \mathbf{P} \tag{6.70}$$

It is straightforward to verify that this generalized inverse is dynamically consistent: we can apply (6.70) to (6.7), and then subsequently to (6.69), to see that $\ddot{\mathbf{x}} = \mathbf{0}$ for any τ_0. Also because $\mathbf{M}_c^{-1} \mathbf{P}$ is unique, the dynamically consistent inverse is also unique. Additionally, this generalized inverse minimizes the instantaneous kinetic energy in the constrained space.† For notational simplicity, we define $\mathbf{N} = \mathbf{I} - \mathbf{J}^T \mathbf{J}^{T\#}$ and write the control equation as

$$\tau = \mathbf{J}^T \mathbf{F} + \mathbf{N} \tau_0 \tag{6.71}$$

6.5.2 OPERATIONAL-SPACE CONTROL WITH UNDERACTUATION

When the system contains passive degrees-of-freedom ($l > 0$), we may no longer be able to generate desired end-effector forces using a direct mapping to joint torques via a Jacobian transpose. The torques we can generate are limited by the following constraint which always must be satisfied,

$$\tau = \mathbf{B} \tau \tag{6.72}$$

Although in general $\mathbf{J}^T \mathbf{F} \neq \mathbf{BJ}^T \mathbf{F}$, we may still be able to satisfy (6.72) by adding a null-space component. When doing so, we can still guarantee that the additional null-space torque will not affect the task-space dynamics. Note, however, because we cannot compensate for task-space dynamics without the addition of null-space torques, we are no longer able to decouple task- and null-space dynamics. Applying (6.71) to (6.72) results in

$$\mathbf{J}^T \mathbf{F} + \mathbf{N} \tau_0 = \mathbf{BJ}^T \mathbf{F} + \mathbf{BN} \tau_0 \tag{6.73}$$

and we can solve for τ_0:

$$(\mathbf{I} - \mathbf{B}) \mathbf{J}^T \mathbf{F} = -(\mathbf{I} - \mathbf{B}) \mathbf{N} \tau_0 \tag{6.74}$$

$$\tau_0 = -\left[(\mathbf{I} - \mathbf{B}) \mathbf{N} \right]^+ (\mathbf{I} - \mathbf{B}) \mathbf{J}^T \mathbf{F} \tag{6.75}$$

---

* For a proof, please see [16].
† Again for a proof, see [16].

again using the Moore–Penrose pseudo-inverse. Provided (6.74) has at least one valid solution for τ_0, we can use (6.75) in (6.71) and write the control equation as

$$\tau = \left(\mathbf{I} - \mathbf{N}\left[(\mathbf{I} - \mathbf{B})\,\mathbf{N}\right]^+ (\mathbf{I} - \mathbf{B})\right) \mathbf{J}^T \mathbf{F} \qquad (6.76)$$

Also, because $(\mathbf{I} - \mathbf{B})$ is an orthogonal projection, the equation simplifies to:

$$\tau = \left(\mathbf{I} - \mathbf{N}\left[(\mathbf{I} - \mathbf{B})\,\mathbf{N}\right]^+\right) \mathbf{J}^T \mathbf{F} \qquad (6.77)$$

If the system has sufficient redundancy, there may be an infinite number of possible control solutions. However, by using the Moore–Penrose pseudo-inverse in (6.75), we are computing the minimum possible $\|\tau_0\|$. Thus (6.77) represents the operational-space control solution with the minimum possible null-space effect. We consider other possible solutions in the following sections.

Note that (6.77) is a general equation that can be applied to both unconstrained and constrained systems. In the unconstrained case, for example, a humanoid robot jumping in the air, the controller generates dynamically consistent null-space motion in order to compensate for lost torque at the passive joints. However, in the constrained case, the controller uses a combination of both motion and constraint forces to generate torque at the passive joints.

If we have more than one task, we can address underactuation using dynamically consistent torque as above, but the torque must be dynamically consistent with all tasks. As an example, consider two tasks with task 1 having a higher priority than task 2. We define the augmented Jacobian of all tasks as $\mathbf{J}_a^T = \begin{bmatrix} \mathbf{J}_1^T & \mathbf{J}_2^T \end{bmatrix}$, and the null space projector that is dynamically consistent with all tasks as $\mathbf{N}_a = \mathbf{I} - \mathbf{J}_a^T \mathbf{J}_a^{T\#}$. Then we write the operational-space control equation as

$$\tau = \mathbf{J}_1^T \mathbf{F}_1 + \mathbf{N}_1 \mathbf{J}_2^T \mathbf{F}_2 + \mathbf{N}_a \tau_0 \qquad (6.78)$$

Solving for τ_0 as done previously, we can write the control solution as

$$\tau = \left(\mathbf{I} - \mathbf{N}_a\left[(\mathbf{I} - \mathbf{B})\,\mathbf{N}_a\right]^+\right) \left(\mathbf{J}_1^T \mathbf{F}_1 + \mathbf{N}_1 \mathbf{J}_2^T \mathbf{F}_2\right) \qquad (6.79)$$

The multitask solution above assumes we have sufficient redundancy remaining after the assignment of all tasks. If all DOFs of the robot are accounted for in $\mathbf{J}_a$ (e.g., Rank$(\mathbf{J}_a) = n$), then it is impossible to add any motion that will not conflict with at least one task. However, if the robot has constraints, we can still add constraint forces to resolve underactuation without inducing any additional joint-space motion. We use the equation:

$$\tau = \mathbf{J}^T \mathbf{F} + (\mathbf{I} - \mathbf{P})\,\tau_C \qquad (6.80)$$

where we have replaced $\mathbf{N}\tau_0$ of (6.71) with $(\mathbf{I} - \mathbf{P})\,\tau_C$, the torques that induce only constraint forces and zero joint acceleration. The replacement is legitimate because $(\mathbf{I} - \mathbf{P})$ projects into a subspace of $\mathbf{N}$; that is, $(\mathbf{I} - \mathbf{P}) = \mathbf{N}(\mathbf{I} - \mathbf{P})$. Then we can satisfy (6.72) with the following control equation. This equation is derived similarly to (6.76), but replacing $\mathbf{N}$ with $(\mathbf{I} - \mathbf{P})$:

$$\tau = \left(\mathbf{I} - (\mathbf{I} - \mathbf{P})\left[(\mathbf{I} - \mathbf{B})(\mathbf{I} - \mathbf{P})\right]^+ (\mathbf{I} - \mathbf{B})\right) \mathbf{J}^T \mathbf{F} \qquad (6.81)$$

As both $(\mathbf{I} - \mathbf{P})$ and $(\mathbf{I} - \mathbf{B})$ are orthogonal projections, the equation simplifies to:

$$\tau = \left(\mathbf{I} - [(\mathbf{I} - \mathbf{B})(\mathbf{I} - \mathbf{P})]^+\right)\mathbf{J}^T\mathbf{F} \tag{6.82}$$

and even further reduces to:

$$\tau = [\mathbf{PB}]^+ \mathbf{J}^T \mathbf{F} \tag{6.83}$$

The above solution is significant for a number of reasons. First, it will produce identical joint acceleration as if the system were fully actuated (but constrained). The null-space component, required to resolve underactuation, produces only constraint force and no additional motion. Therefore this controller is the solution that resolves underactuation with a minimal amount of kinetic energy. Second, the projector $[\mathbf{PB}]^+$ is identical to the projector of constrained underactuated inverse dynamics controllers. For example, to achieve a desired joint-space acceleration $\ddot{\mathbf{q}}_d$, provided $\ddot{\mathbf{q}}_d$ is constraint consistent, we can use the following controller independent of constraint forces [12],

$$\tau = [\mathbf{PB}]^+ (\mathbf{M}\ddot{\mathbf{q}}_d + \mathbf{h}) \tag{6.84}$$

Finally, (6.83) is mathematically identical to the operational space controller derived in works by Sentis et al. for task-space motion control of humanoid and legged robots, but in a much reduced form. For example, the inertia matrix only appears in $\mathbf{F}$ and is not used in the projector or Jacobian as in Sentis [24]. Note that this point is more thoroughly developed in Righetti et al. [20], which specifically shows a proof of equivalence between orthogonal and oblique projector operators in the context of joint acceleration or inverse dynamics control.

Because the previous controller (6.83) only uses constraint forces to resolve underactuation, for comparison, we can consider a controller that only uses null-space motion. The controller is written as

$$\tau = \left(\mathbf{I} - \mathbf{PN}\,[(\mathbf{I} - \mathbf{B})\,\mathbf{PN}]^+\right)\mathbf{J}^T\mathbf{F} \tag{6.85}$$

where we have replaced $\mathbf{N}$ of (6.77) with $\mathbf{PN}$. Again, this is possible because $\mathbf{PN} = \mathbf{NPN}$. As this controller only uses null-space motion to resolve underactuation, it attempts not to change the constraint forces generated by task dynamics.

6.6 APPENDICES

6.6.1 Exoskeleton Control Architecture

The control architecture consists of independent PD servocontrollers at each joint (implemented on individual Sarcos Advanced Joint Controller analog circuit boards), with additional feedforward torque commands computed on a centralized controller running on two Motorola PPC 603 parallel processors with the commercial real-time operating system vxWorks (Windriver Systems). Potentiometers and load cells at each joint are sampled at 960 Hz to provide positional and torque feedback, respectively. Joint velocities and accelerations are computed numerically by differentiating the position signal. The signals are filtered with a second-order Butterworth filter with cut-off frequency of 33.6 Hz for position, velocity, and torque and 4.8 Hz for acceleration. The acceleration signal requires more aggressive filtering because of the noise amplified by the numerical differentiation. The centralized controller updates the feedforward commands and PD set points at 480 Hz.

6.6.2 EXOSKELETON CONTROL FORMULATION

Ideally, the user should not be burdened (unintentionally) by the exoskeleton while executing movements. Therefore the control scheme needs to compensate for the exoskeleton's gravity, Coriolis, and inertia forces. Typically, robot manipulanda use force sensors at the end-effector to measure the forces applied by the user [8], and apply control laws that generate a particular impedance characteristic at the point where the user holds the manipulandum. In contrast to these previous approaches, which focused on investigations of the end-effector movement of human arms, our current robotic setup is to examine joint-level effects of human motor control. Thus, we do not have a constraint in the form of a desired impedance at the end-effector, but rather impedance control for every DOF. Inasmuch as we cannot reliably attach force sensors between the human arm and every link of the robot, we resorted to a model-based control approach.

6.6.3 LOW-LEVEL JOINT CONTROL

The low-level joint controllers are governed by the equation

$$\mathbf{u} = \mathbf{K}_D(\dot{\mathbf{q}}_d - \dot{\mathbf{q}}) + \mathbf{K}_P(\mathbf{q}_d - \mathbf{q}) + \mathbf{u}_{ff} \tag{6.86}$$

where $\mathbf{u} \in \mathbb{R}^7$ is the vector of motor command torques, $\mathbf{q}, \dot{\mathbf{q}} \in \mathbb{R}^7$ are the vectors of joint position and velocity, and $\mathbf{K}_D, \mathbf{K}_P$ are diagonal gain matrices. The desired joint position and velocity vectors $\mathbf{q}_d, \dot{\mathbf{q}}_d \in \mathbb{R}^7$ and the feedforward torque command $\mathbf{u}_{ff} \in \mathbb{R}^7$ are set by the centralized controller. Ideally, we would set $\dot{\mathbf{q}}_d = \dot{\mathbf{q}}$ and $\mathbf{q}_d = \mathbf{q}$ in this controller, and compute an inverse dynamics-based feedforward command $\mathbf{u}_{ff}$ to eliminate inertial, Coriolis, and gravity forces. However, as such a controller is neutrally stable and difficult to realize in light of inevitable modeling errors in the dynamics model, a more prudent control approach is required. In order to keep a small amount of a position reference for enhanced stability, we define

$$\mathbf{q}_d^{n+1} = \mathbf{q}_d^n + \epsilon(\mathbf{q}^n - \mathbf{q}_d^n) \tag{6.87}$$

where n is the discrete timestep of the control loop, and ϵ is a constant factor between 0 and 1. Thus $\mathbf{q}_d$ is a filtered version of $\mathbf{q}$, and removes high-frequency noise at the cost of a lag behind the current joint state. The filter parameter ϵ is chosen high enough such that speed of typical human arm movements is well within the bandwidth of the filter. If we choose $\epsilon = 1$, we effectively achieve the setting $\mathbf{q}_d = \mathbf{q}$. We indeed set the desired velocity $\dot{\mathbf{q}}_d = \dot{\mathbf{q}}$, but we maintain a small amount of damping in the feed–forward command $\mathbf{u}_{ff}$.

6.6.4 GRAVITY, CORIOLIS, AND INERTIA COMPENSATION

We assume the exoskeleton is governed by the well-known rigid-body dynamics model of a manipulator robot arm given by the equation

$$\mathbf{M}(\mathbf{q})\ddot{\mathbf{q}} + \mathbf{C}(\mathbf{q}, \dot{\mathbf{q}}) + \mathbf{G}(\mathbf{q}) = \mathbf{u} \tag{6.88}$$

where $\mathbf{M}(\mathbf{q}) \in \mathbb{R}^{7 \times 7}$ is the mass or inertia matrix, $\mathbf{C}(\mathbf{q}, \dot{\mathbf{q}}) \in \mathbb{R}^7$ denotes centrifugal and Coriolis forces, and $\mathbf{G}(\mathbf{q}) \in \mathbb{R}^7$ denotes the gravity force [22]. Theoretically, we can completely cancel the dynamics of the exoskeleton robot with the feedforward control law,

$$\mathbf{u}_{ff} = \tilde{\mathbf{M}}(\mathbf{q})(\ddot{\mathbf{q}} - \mathbf{K}'_D \dot{\mathbf{q}}) + \tilde{\mathbf{C}}(\mathbf{q}, \dot{\mathbf{q}}) + \tilde{\mathbf{G}}(\mathbf{q}) \tag{6.89}$$

where $\tilde{\mathbf{M}}, \tilde{\mathbf{C}}, \tilde{\mathbf{G}}$ are our estimates of the robot's inertia, Coriolis, and gravity matrices, and $\mathbf{K}'_D$ is a diagonal matrix of damping gains. Assuming perfect parameter estimation, the robot's dynamics will be eliminated leaving only the damping term $\tilde{\mathbf{M}}(\mathbf{q})\mathbf{K}'_D \dot{\mathbf{q}}$ which is required to maintain stability. The damping gains can be individually tuned for each joint to apply the appropriate impedance characteristic at each joint. Small gains can be used to keep the user's motion as unconstrained as possible. Premultiplying the damping term by the inertia matrix has the consequence of amplifying any modeling inaccuracies of the inertia matrix. Therefore, we move the damping term outside the inertia matrix product, which empirically we find provides a greater margin of stability. Combining this control law with Equation (6.86) results in the equation

$$\mathbf{u} = \tilde{\mathbf{M}}(\mathbf{q})\ddot{\mathbf{q}} + \tilde{\mathbf{C}}(\mathbf{q}, \dot{\mathbf{q}}) + \tilde{\mathbf{G}}(\mathbf{q}) - \mathbf{K}'_D \dot{\mathbf{q}} + \mathbf{K}_P(\mathbf{q}_d - \mathbf{q}) \tag{6.90}$$

with $\mathbf{q}_d$ defined by Equation (6.87).

6.6.5 PARAMETER IDENTIFICATION

The estimates of our model parameters, $\tilde{\mathbf{M}}, \tilde{\mathbf{C}}, \tilde{\mathbf{G}}$, are determined with system identification techniques [2]. We recorded an hour of robot trajectories in response to sufficiently exciting desired trajectories of pseudo-random motor commands (including sine waves of various frequencies at the joint level and discrete endpoint movements at various speeds). These data were used to regress the rigid-body dynamics parameters acting on each DOF. However, we noticed that due to unmodeled nonlinearities of the robot, we obtained partially physically inconsistent rigid-body parameters, for example, nonpositive definite link inertia matrices and negative viscous friction coefficients. We improved this estimation with a novel nonlinear rigid-body dynamics identification algorithm that guaranteed a physically correct rigid-body model [18]. This model significantly increased the stability of our control system.

REFERENCES

1. Aghili, F. A unified approach for inverse and direct dynamics of constrained multibody systems based on linear projection operator: Applications to control and simulation. *IEEE Trans. Robot.* **21**(5):834–849, Oct 2005.

2. An, C.H., Atkeson, C., and Hollerbach, J. *Model-Based Control of a Robot Manipulator.* *Cambridge, MA: MIT Press*, 1988.

3. Baerlocher, P. and Boulic, R. Task-priority formulations for the kinematic control of highly redundant articulated structures. *IROS 1998*, **1**:323–329, 1998.

4. Bernstein, N. *The Co-Ordination and Regulation of Movements. London: Pergamon Press*, 1967.

5. Chiaverini, S. Singularity-robust task-priority redundancy resolution for real-time kinematic control of robot manipulators. *IEEE Trans. Robot. Autom.* **13**(3):398–410, 1997.

6. Conditt, M., Gandolfo, F., and Mussa-Ivaldi, F.A. The motor system does not learn the dynamics of the arm by rote memorization of past experience. *J. Neurophysiol*, **78**(1):554–560, 1997.

7. Gandolfo, F., Mussa-Ivaldi, F.A., and Bizzi, E. Motor learning by field approximation. *Proc. Natl. Acad. Sci. USA*, **93**(9):3843–3846, 1996.

8. Gomi, H. and Kawato. Equilibrium-point control hypothesis examined by measured arm stiffness during multijoint movement. *Science*, **272**(5258):117–20, Apr 1996.

9. Khatib, O. A unified approach for motion and force control of robot manipulators: The operational space formulation. *IEEE J. Robot. Automati.* **3**(1):43–53, 1987.

10. Lashley, K.S. Basic neural mechanisms in behavior. *Psychol. Rev.* **37**:1–24, 1930.

11. Lockhart, D.B. and Ting, L.H. Optimal sensorimotor transformations for balance. *Nat. Neurosci.* **10**(10):1329–36, Oct 2007.

12. Mistry, M., Buchli, J., and Schaal, S. Inverse dynamics control of floating base systems using orthogonal decomposition. In *Proceedings of the 2010 IEEE Int. Conference on Robotics and Automation*, pp. 3406–3412, 2010.

13. Mistry, M., Mohajerian, P., and Schaal, S. Arm movement experiments with joint space force fields using an exoskeleton robot. In *9th International Conference on Rehabilitation Robotics*, pp. 408–413, January 2005.

14. Mistry, M., Mohajerian, P., and Schaal, S. An exoskeleton robot for human arm movement study. In *2005 IEEE/RSJ International Conference on Intelligent Robots and Systems*, pp. 4071–4076, July 2005.

15. Mistry, M., Nakanishi, J., and Schaal, S. Task space control with prioritization for balance and locomotion. *IROS 2007*, pp. 331–338, January 2007.

16. Mistry, M. and Righetti, L. Operational space control of constrained and underactuated systems. In *2011 Robotics: Science and Systems Conference*, Los Angeles, June 2011.

17. Nakamura, Y., Hanafusa, H., and Yoshikawa, T. Task-priority based redundancy control of robot manipulators. *Int. J. Robot. Res.* **6**(2):3–15, June 1987.

18. Nakanishi, J., Cory, R., Mistry, M., Peters, J., and Schaal, S. Operational space control: A theoretical and empirical comparison. *Int. J. Robot. Res.* **27**(6):737–757, December 2008.

19. Peters, J., Mistry, M., Udwadia, F., Nakanishi, J., and Schaal, S. A unifying framework for robot control with redundant dofs. *Auton. Robots.* **24**(1):1–12, December 2008.

20. Righetti, L., Buchli, J., Mistry, M., and Schaal, S. Inverse dynamics control of floating-base robots with external constraints: a unified view. In *Proceedings of the 2011 IEEE International Conference on Robotics and Automation*, 2011.

21. Scholz, J.P. and Schöner, G. The uncontrolled manifold concept: Identifying control variables for a functional task. *Exp. Brain. Res.* **126**(3):289–306, June 1999.

22. Sciavicco, L. and Siciliano, B. *Modeling and Control of Robot Manipulators*, 2nd ed. 2001.

23. Sentis, L. and Khatib, O. A whole-body control framework for humanoids operating in human environments. In *Proceedings of the 2006 IEEE International Conference on Robotics and Automation*, pp. 2641–2648, 2006.

24. Sentis, L. *Synthesis and Control of Whole-Body Behaviors in Humanoid Systems*. PhD thesis, Stanford University, Stanford, CA, June 2007.

25. Shadmehr, R. and Mussa-Ivaldi, F.A. Adaptive representation of dynamics during learning of a motor task. *J. Neurosci.* **14**(5 Pt 2):3208–3224, 1994.

26. Siciliano, B. and Slotine, J.-J. A general framework for managing multiple tasks in highly redundant robotic systems. In *Fifth International Conference on Advanced Robotics*, Vol. 2, pp. 1211–1216, 1991.

27. Singh, K. and Scott, S.H. A motor learning strategy reflects neural circuitry for limb control. *Nat. Neurosci.* **6**(4):399–403, April 2003.

28. Soechting, J.F., Buneo, C.A., Herrmann, U., and Flanders, M. Moving effortlessly in three dimensions: does Donders' law apply to arm movement? *J. Neurosci.* **15**(9):6271–80, September 1995.

29. Strang, G. *Linear Algebra and Its Applications*, 3rd ed. 1988.

30. Todorov, E. and Jordan, M.I. Optimal feedback control as a theory of motor coordination. *Nat. Neurosci.* **5**(11):1226–35, October 2002.

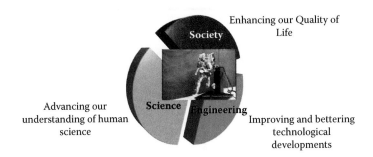

Enhancing our Quality of
Life

Advancing our
understanding of human
science

Improving and bettering
technological
developments

FIGURE 1.1   Research paradigm: Science, engineering, and society.

FIGURE 1.9   The ETL-Humanoid system. (Courtesy of Yasuo Kuniyoshi. With permission.)

FIGURE 1.11   The iCub humanoid robot (iCub at the TUM-ICS lab).

FIGURE 1.22   Visual attention. The robot actively saccades to visually acute movement.

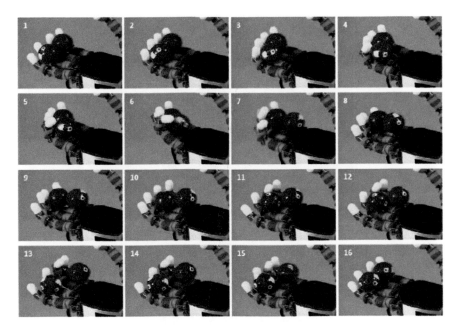

FIGURE 2.3 Frames representing the ball-swapping skill obtained using the human senso-rimotor learning paradigm. The numbers in the top left corner of the images indicate the chronological order of the frames. (Adopted from Moore and Oztop 2012. With permission.)

FIGURE 3.6 Preshape-based grasping of daily objects with Shadow Hand. (From Röthling et al. 2007. With permission.)

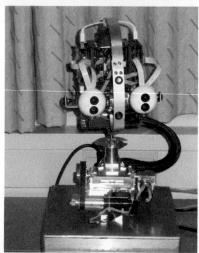

FIGURE 5.2 Heads of two humanoid robots. The two robots in the left and central images have foveal cameras mounted above peripheral cameras, and the robot in the right image has foveal cameras on the outer side of the peripheral cameras.

(a)

(b)

(c)

FIGURE 7.2 (a) Simple 3D biped simulation model of the CBi. The biped model has 10 degrees-of-freedom; height: 1.59 m, total weight: 95 kg. (b) Human-sized humanoid robot CBi [7]. (c) Small humanoid robot used in the experiment.

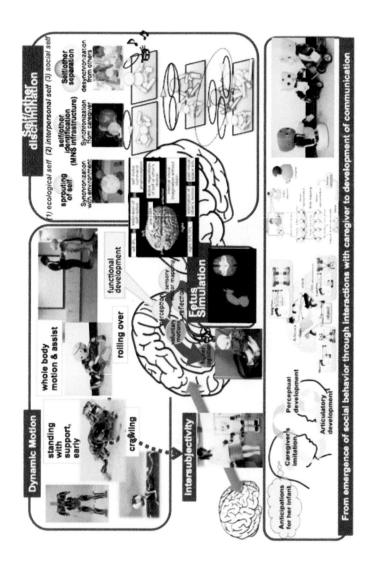

FIGURE 10.2 An overview of the achievements by the JST ERATO Asada Project with some new projects.

7 Humanoid Locomotion and the Brain

Jun Morimoto

CONTENTS

7.1   INTRODUCTION

In this chapter, we introduce a two-layered biologically inspired biped learning system composed of a lower-level central pattern generator (CPG) and an upper-level reinforcement learning (RL) module. In this proposed approach, the lower-level CPG is used to generate nominal periodic walking patterns and also used to synchronize the periodic patterns with sensory inputs in order to construct an attractive limit cycle. On the other hand, the upper-level reinforcement learning module is used to modulate parameters of the CPG in order to improve walking performances with respect to the provided objective function. Figure 7.1 shows the schematic diagram of the biped learning system.

Biological systems seem to have a simpler but more robust locomotion strategy [26] than existing biped walking controllers for humanoid robots (e.g., [2]). For example, Grillner [17] and [39] showed that the cat locomotion system can generate a walking pattern without using higher brain functions. An early study of biologically inspired

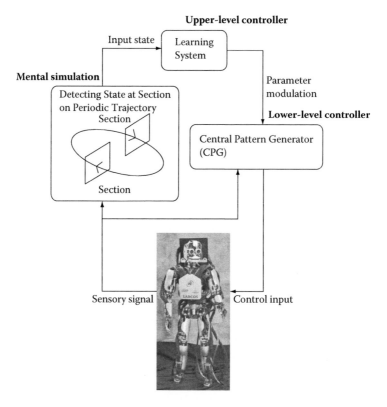

FIGURE 7.1   Schematic diagram of our biped learning scheme. The central pattern generator (CPG) generates periodic patterns for biped locomotion. Detected sensory signals are sent to the CPG and also to the learning system. The learning system updates the policy and outputs the modulation command to the CPG only when the periodic trajectory passes through the defined Poincaré sections.

approaches to bipedal locomotion [45] suggested that the synchronization property of the neural system with periodic sensor inputs plays an important role for robust locomotion control.

After these leading studies, there is growing interest in biologically inspired locomotion control utilizing CPG modeled by coupled neural oscillators [12,14,15,29] or using a phase oscillator model with phase reset methods [36,47]. These studies make use of foot contact information or ground reaction forces in exploiting the entrainment property of the neural or phase oscillator model.

However, to adapt CPG to new environments, the parameters of the CPG need to be modified. To cope with this environmental change, in the proposed approach, we use a mental-simulation–based (or model-based) RL method [10,30,44] in which the learning system only interacts with a mentally simulated model. In this model-based approach, samples to improve parameters of policies can be generated from the simulated model without directly interacting with the real environment. Accordingly, once the environment model is properly identified, policy parameters can be improved without using real biped robots.

On the other hand, the correct robot model and the ground contact model are difficult to identify. Therefore, we consider using a Poincaré map, which has been used to evaluate local stability of periodic patterns, as the mental simulation model and representing the Poincaré map by using a nonparametric function approximation method.

In the following sections, we introduce the two-layered biped learning system in detail. In Section 7.2, an implementation of the CPG model is introduced. Experimental results show that a biped robot model and a real humanoid robot are able to walk successfully without using carefully designed walking patterns. In Section 7.3, the model-based RL approach using Poincaré map as the mentally simulated environment is introduced. Walking patterns generated by the CPG are successfully improved by using the proposed learning approach.

7.2   LOWER-LEVEL PERIODIC PATTERN GENERATOR

As presented in Figure 7.1, we consider using a CPG model as the lower-level controller of the proposed framework. Here, we introduce our CPG model composed of coupled phase oscillators.

Several studies designed walking trajectories as a function of a physical variable of the robot (e.g., ankle joint angle) [9,20,48]. However, a neural oscillator model has complex dynamics and many parameters to be selected [29,45]. Other approaches that have synchronization mechanisms usually require proper gait design [9,36,47,48].

In the proposed approach, we undertake the development of a simple biped controller by means of a coupled oscillator system, which is said to exist in vertebrates and is widely referred to as a central pattern generator [18].

Many biped walking studies have emphasized that humanoid robots have inverted pendulum dynamics, with the top at the center of mass and the base at the center of pressure, and proposed control strategies to stabilize the dynamics [19,22,28,35,41].

We propose using the center of pressure to detect the phase of the inverted pendulum dynamics. (1) We use simple periodic functions (sinusoids) as the desired joint trajectories. (2) We show that synchronization of the desired trajectories at each joint with the inverted pendulum dynamics can generate stepping and walking. (3) Our nominal gait patterns are sinusoids, therefore our approach does not need careful design of desired gait trajectories.

We apply an oscillator model to a simulated biped model and to a human-sized humanoid robot CBi [7] (see Figure 7.2) for biped walking in a real environment. First, we introduce our biologically inspired biped locomotion strategy, which uses modulated sinusoidal patterns via a coupled oscillator model, described in Sections 7.2.1 and 7.2.2. In Section 7.2.3, we apply our proposed approach to the simulated robot model and also show our experimental results (see Figures 7.2a and b).

7.2.1 MODULATION OF SINUSOIDAL PATTERNS BY A COUPLED OSCILLATOR MODEL

Our biped control approach uses a coupled phase oscillator model [40] to modulate sinusoidal patterns. The aim of using the coupled phase oscillator model is to synchronize periodic patterns generated by the controller with the dynamics of the robot.

We also show a strategy to design a nominal desired joint angle. One of the simplest ways to generate a periodic pattern at each joint is using only one sinusoidal basis function to represent the desired joint angle. By only using one sinusoidal basis function at each joint, we have the smallest numbers of parameters to represent periodic patterns at each joint. We introduce our stepping and walking controllers that use the desired joint angle represented by the sinusoidal function.

7.2.1.1 Coupled Oscillator Model

Here, we consider the behavior of the following dynamics of the phase of a biped controller ϕ_c and the phase of the robot dynamics ϕ_r,

$$\dot{\phi}_c = \omega_c + K_c \sin(\phi_r - \phi_c), \tag{7.1}$$

$$\dot{\phi}_r = \omega_r + K_r \sin(\phi_c - \phi_r), \tag{7.2}$$

where $\omega_c > 0$ and $\omega_r > 0$ are natural frequencies of the controller and the robot dynamics, and K_c, K_r are positive coupling constants. We can find two fixed points if $|\omega_c - \omega_r| < K_c + K_r$. There is no fixed point if $|\omega_c - \omega_r| > K_c + K_r$. A saddle-node bifurcation occurs when $|\omega_c - \omega_r| = K_c + K_r$.

If $|\omega_c - \omega_r| < K_c + K_r$, the oscillators run with the phase difference: $\Psi^* = \phi_r - \phi_c = \sin^{-1}((\omega_r - \omega_c)/(K_c + K_r))$ and the compromise frequency: $\omega^* = (K_r \omega_c + K_c \omega_r)/(K_c + K_r)$ when they are entrained [40].

We showed the typical time profile of phase difference $\Phi = \phi_c - \phi_r$, and angular frequencies ω_c and ω_r in Figure 7.3. Although usually the biped dynamics cannot be represented by the simple phase dynamics (7.2), we can still detect the phase from the robot dynamics as described in Section 7.2.1. Then, we use (7.1) to adjust the controller phase under the assumption that the phase dynamics detected from the biped dynamics keeps a similar property to (7.2).

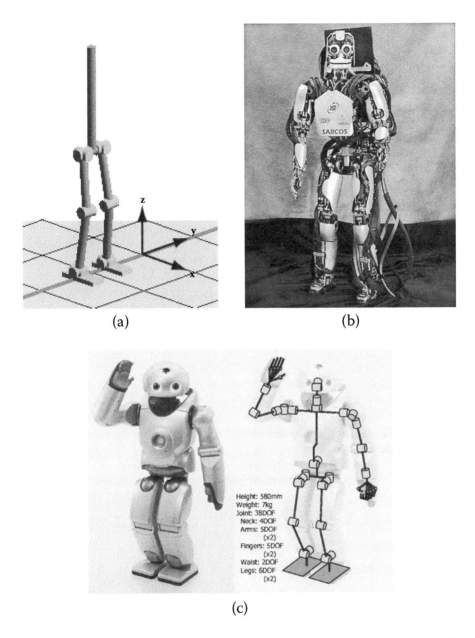

(a)

(b)

(c)

FIGURE 7.2 (a) Simple 3D biped simulation model of the CBi. The biped model has 10 degrees-of-freedom; height: 1.59 m, total weight: 95 kg. (b) Human-sized humanoid robot CBi [7]. (c) Small humanoid robot used in the experiment. See color insert.

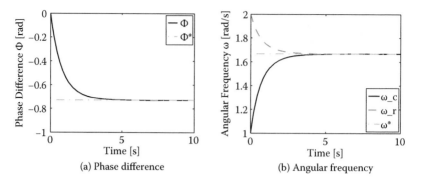

FIGURE 7.3   Typical time profile of the coupled oscillator system. (a) Phase difference $\Psi = \phi_r - \phi_c$. The phase difference converged to the analytically derived value Ψ^* depicted by dash-dot line. (b) Angular frequencies ω_r and ω_c. Angular frequency of each oscillator converged to the analytically derived compromise frequency ω^*.

7.2.1.2 Phase Detection from the Robot Dynamics

As previous studies have pointed out, controlling the inverted pendulum dynamics represented by the center of mass and the center of pressure (Figure 7.4a) is a major issue in controlling biped robots. We consider the inverted pendulum dynamics on a lateral plane that has a four-dimensional state space $\mathbf{x} = (y, \dot{y}, \psi^{roll}, \dot{\psi}^{roll})$, depicted in Figure 7.4a.

To detect the phase from the inverted pendulum dynamics, we project the four-dimensional state space to a two-dimensional state space. Then, we convert the two-dimensional state space to the phase space.

In this study, we consider the center of pressure y and the velocity of the center of pressure $\dot{y}$ as the variables in the two-dimensional state space because detecting these values by force sensors on soles is easy for real robots. Therefore, we detect the phase as

$$\phi(\mathbf{y}_{cop}) = -\arctan\left(\frac{\dot{y}_{cop}}{y_{cop}}\right) \tag{7.3}$$

where $\mathbf{y}_{cop} = (y_{cop}, \dot{y}_{cop})$ (see Figure 7.4).

7.2.1.3 Simplified COP Detection

The center of pressure depends on a coordinate system, and we need a kinematic model to detect COP. Alternatively, we use an approximate center of pressure:

$$y = \frac{y^l_{foot} F^l_z + y^r_{foot} F^r_z}{F^l_z + F^r_z} \tag{7.4}$$

where F^l_z and F^r_z represent the left and right ground reaction force, respectively, and y^l_{foot} and y^r_{foot} are the lateral position of each foot. We assume that the feet are symmetrically placed $-y^l_{foot} = y^r_{foot}$. Because we only use this center of pressure to

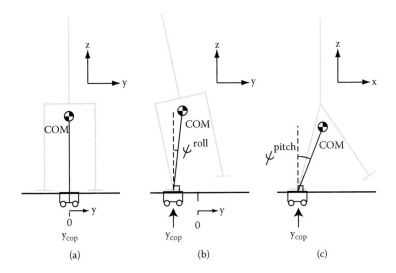

FIGURE 7.4 Inverted pendulum model represented by the center of pressure (COP) and the center of mass (COM). ψ^{roll} denotes the roll angle of the pendulum. ψ^{pitch} denotes the pitch angle of the pendulum. y_{cop} denotes the center of pressure.

detect the phase of the robot dynamics $\phi_r(\mathbf{x})$ in Equation (7.3), the scale of the foot position y^l_{foot} and y^r_{foot} can be arbitrary. We simply set $y^l_{foot} = -y^r_{foot} = 1.0$ m. This simplified COP detection does not require the kinematic model.

7.2.1.4 Phase Coordination

In this approach, we use four oscillators with phases ϕ^i_c, where $i = 1, 2, 3, 4$. We introduce coupling between the oscillators and the phase of the robot dynamics $\phi(y_{cop})$ in (7.3) to regulate the desired phase relationship between the oscillators:

$$\dot{\phi}^i_c = \omega_c + K_c \sin(\phi(y_{cop}) - \phi^i_c + \phi^i_d), \tag{7.5}$$

where ϕ^i_d is the desired phase difference, K_c is a coupling constant, and ω_c is the natural angular frequency of oscillators.

We use four desired phase differences, $\{\phi^1_d, \phi^2_d, \phi^3_d, \phi^4_d\} = \{-\frac{1}{2}\pi, 0.0, \frac{1}{2}\pi, \pi\}$, to make symmetric patterns for the stepping movement with the left and right limbs, and also to make symmetric patterns for forward movement with the left and right limbs. Figure 7.6 shows the phase coordination.

Two parameters $\{\omega_c, K_c\}$ need to be selected to define the phase oscillator dynamics.

We empirically found that the natural frequency of a linear pendulum with the length l can be a good candidate for the natural frequency of the controller as $\omega_c = \sqrt{g/l}$, where g denotes the acceleration due to the gravity and l denotes the COM height when a biped stands straight.

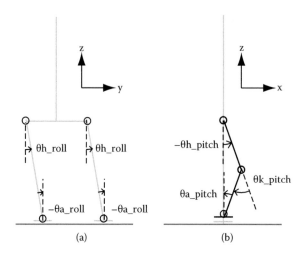

FIGURE 7.5   Joint index of biped model. (a) Roll joints for lateral movements. (b) Pitch joints for lateral and forward movements.

By considering the insight from the oscillator dynamics (7.1) and (7.2), we need to use the sufficiently large coupling constant K_c to satisfy $|\omega_c - \omega_r| < K_c + K_r$ for keeping fixed points.

7.2.2   Desired Joint Angles

7.2.2.1   Roll Joint Movements

First, we introduce a controller to generate side-to-side movement. We control the hip joints $\theta_{hip^{roll}}$ and the ankle joints $\theta_{ankle^{roll}}$ (Figure 7.5a) for this movement. Desired joint angles for each joint are:

$$\theta^d_{hip^{roll}}(\phi_c) = A_{hip^{roll}} \sin(\phi_c) \tag{7.6}$$

$$\theta^d_{ankle^{roll}}(\phi_c) = -A_{ankle^{roll}} \sin(\phi_c) \tag{7.7}$$

where $A_{hip^{roll}}$ and $A_{ankle^{roll}}$ are the amplitudes of a sinusoidal function for side-to-side movements at the hip and the ankle joints, and we use an oscillator with the phase $\phi_c = \phi_c^1$.

7.2.2.2   Pitch Joint Movements

To achieve foot clearance, we generate vertical movement of the feet (Figure 7.5b) by using simple sinusoidal trajectories:

$$\theta^d_{hip^{pitch}}(\phi_c) = A_{pitch} \sin(\phi_c) + \theta^{res}_{hip^{pitch}}$$

$$\theta^d_{knee^{pitch}}(\phi_c) = -2A_{pitch} \sin(\phi_c) + \theta^{res}_{knee^{pitch}}$$

$$\theta^d_{ankle^{pitch}}(\phi_c) = -A_{pitch} \sin(\phi_c) + \theta^{res}_{ankle^{pitch}}, \tag{7.8}$$

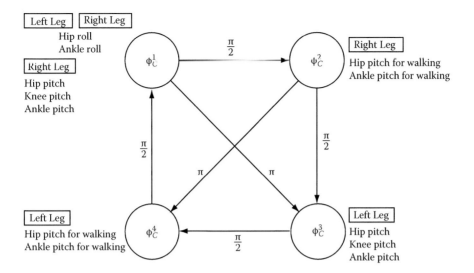

FIGURE 7.6 Phase coordination. We use four oscillators with phases $\{\phi_c^1, \phi_c^2, \phi_c^3, \phi_c^4\}$ to make symmetric patterns for a lateral movement with the left and right limbs, and also to make symmetric patterns for a forward movement with the left and right limbs. Arrows "→" indicate phase advance, for example, $\phi_c^2 = \phi_c^1 + (\pi/2)$ and $\phi_c^3 = \phi_c^1 + \pi$.

where A_{pitch} is the amplitude of a sinusoidal function to achieve foot clearance; $\theta_{hip^{pitch}}^{res}, \theta_{knee^{pitch}}^{res}, \theta_{ankle^{pitch}}^{res}$ represent the rest posture of the hip, knee, and ankle joints, respectively. We use the oscillator with phase $\phi_c = \phi_c^1$ for right limb movement and use the oscillator with phase $\phi_c = \phi_c^3$, which has a phase difference of $\phi_c^3 = \phi_c^1 + \pi$, for left limb movement as in Figure 7.6.

7.2.2.3 Control Input for Walking Behaviors

To walk forward, we use an additional sinusoidal trajectory. Thus, the desired nominal trajectories for the right hip and ankle pitch joints become:

$$\theta_{hip^{pitch}}^{d} \leftarrow A_{hip}^{walk} \sin(\phi_c^2) + \theta_{hip^{pitch}}^{d}(\phi_c^1)$$

$$\theta_{ankle^{pitch}}^{d} \leftarrow -A_{ankle}^{walk} \sin(\phi_c^2) - \theta_{ankle^{pitch}}^{d}(\phi_c^1) \tag{7.9}$$

We use the phase $\phi_c = \phi_c^2$, which has a $\frac{1}{2}\pi$ phase difference of ϕ_c^1 for the right limb. We use the phase $\phi_c = \phi_c^4$, which has a π phase difference with ϕ_c^2. We use ϕ_c^3 and ϕ_c^4 for the left limb instead of ϕ_c^1 and ϕ_c^2.

7.2.3 SIMULATION AND EXPERIMENTAL RESULT

We applied our proposed method to a simple 3D biped robot model (Figure 7.2a), our human-sized humanoid robot CB-i (Figure 7.2b).

7.2.3.1 Simulation Setups

Here we describe our simulation setups. To follow the desired trajectories, the torque output at each joint is given by a PD servo controller:

$$\tau = \mathbf{K}_p(\theta^d(\phi_c) - \theta) + \mathbf{K}_d(\dot{\theta}^d(\phi_c) - \dot{\theta}), \tag{7.10}$$

where $\theta^d(\phi_c) \in \mathbf{R}^{10}$ is the target joint angle vector, $\mathbf{K}_p$ denotes the position gain matrix, and $\mathbf{K}_d$ denotes the velocity gain matrix. Each element of the diagonal position gain matrix $\mathbf{K}_p$ is set to 3,000 and each element of the diagonal velocity gain matrix $\mathbf{K}_d$ is set to 100.

We used the fourth-order Runge–Kutta method with a timestep of $\Delta t = 0.0003$ s to numerically integrate the biped dynamics. The vertical ground reaction force f_z is simulated by a spring-dumper model:

$$f_z = -k_p^z z_{cp} - k_d^z \dot{z}_{cp} \tag{7.11}$$

where $k_p^z = 30,000$ is the spring gain, $k_d^z = 1,000$ is the dumper gain, and z_{cp} denotes the vertical position of a contact point.

The ground reaction force for horizontal directions f_x and f_y are simulated by a viscose friction:

$$f_x = -k_d^x \dot{x}_{cp} \tag{7.12}$$

$$f_y = -k_d^y \dot{y}_{cp} \tag{7.13}$$

where $k_d^x = 2,500$ and $k_d^y = 2,500$ are dumper gains. x_{cp} and y_{cp} are horizontal positions of a contact point.

7.2.3.2 Stepping Movement

We applied our proposed method to the biped robot model.

As we proposed in Section 7.2.1, the natural frequency of the controller is set as $\omega_c = \sqrt{g/l} = 3.6$ rad/s, and the coupling constant is set as $K_c = 10.0$. Then, we compared different parameter settings for the amplitudes $A_{hip^{roll}}$ and $A_{ankle^{roll}}$ in (7.6), (7.7) and A_{pitch} in (7.8).

Figure 7.7 shows results of the comparison. Using large A_{pitch} with small $A_{hip^{roll}}$ and $A_{ankle^{roll}}$ results in falling over. On the other hand, using small A_{pitch} with large $A_{hip^{roll}}$ and $A_{ankle^{roll}}$ cannot make the single support phase. By comparing Figures 7.7a and 7.7b, stepping movement that has the smaller pendulum angle ψ^r tends to have a larger stepping period.

A proper combination of the parameters $A_{hip^{roll}} = A_{ankle^{roll}} = 2.5°$ and $A_{pitch} = 5.0°$, which can make a stepping movement without falling over, generated a stepping movement with period 1.4 s. Equivalently, average angular frequency $\dot{\phi}_c^{av} = 1/T(\phi_c(T+t) - \phi_c(t))$ of the stepping movement was $\dot{\phi}_c^{av} = 4.5$ rad/s, where T is a stepping period.

To show how the coupled oscillator model in Equations (7.1) and (7.2) worked with the biped robot model, we tested a different controller with a different natural

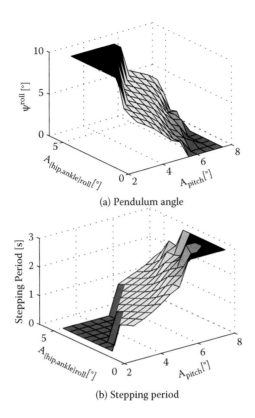

(a) Pendulum angle

(b) Stepping period

FIGURE 7.7   Comparison of using different amplitude parameters. (a) Pendulum angle ψ^{roll} in Figure 7.4a. Region has the value $\psi^{roll} = 10$ representing that the robot cannot make the single support phase with the corresponding parameter selection. Region has the value $\psi^{roll} = 0$ representing that the robot falls over with the corresponding parameter selection. (b) Stepping period. Region has the value 0 s representing that the robot cannot make the single support phase with the corresponding parameter selection. Region has the value 3 s representing that the robot falls over with the corresponding parameter selection.

frequency, $\omega_c = 2.5$ rad/s. Although the natural frequency was different, the modulated averaged frequency $\dot{\phi}_c^{av} = 3.9$ rad/s was much closer to the previous averaged frequency, $\dot{\phi}_c^{av} = 4.5$ rad/s than the selected natural frequency $\omega_c = 2.5$ rad/s.

By considering the compromise frequency ω^* introduced in Section 7.2.1, this result indicates that the current coupling constant $K_c = 10.0$ is large enough to make the controller frequency close to the natural frequency of the robot dynamics.

Figure 7.8 shows trajectories of desired and actual hip joint angles $\theta_{hip^{roll}}^d$, $\theta_{hip^{roll}}$ (see Figure 7.4b). Figure 7.8a represents the result without using a coupled oscillator model. The desired trajectory is the original simple sinusoidal trajectory.

Figure 7.8b represents the result of using a coupled oscillator model. This modulated trajectory made stepping movement possible. Large tracking errors appeared during the single support phase due to not using a very large servo gain. This result shows that our proposed method does not require accurate tracking performance. The

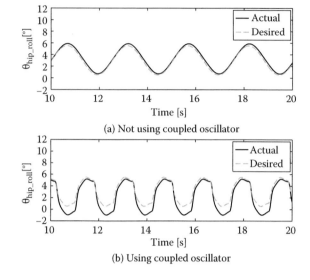

FIGURE 7.8 Generated desired and actual trajectory at hip-roll joint. (a) Not using coupled oscillator; (b) using coupled oscillator. Large tracking error appeared during single support phase due to not using very large servo gain. The result shows that our proposed method does not require accurate tracking performance.

desired trajectory is modulated by the coupling with the phase of the robot dynamics $\phi_r(\mathbf{x})$.

7.2.3.3 Biped Walking

First, we applied our proposed method to generate walking movements by using the simulated biped model. The same parameters as the stepping controller for the natural frequency of the controller $\omega_c = 3.6$ rad/s, and the coupling constant $K_c = 10.0$ are used. We empirically figured out proper amplitude parameters $A_{hip^{roll}} = A_{ankle^{roll}} = 2.5°$ and $A_{pitch} = 6.0°$ for the walking task. We compared different amplitude parameters A_s by setting $A_{hip}^{walk} = A_s$ and $A_{ankle}^{walk} = A_s/2$ in (7.9).

Figure 7.9 shows the results of the comparison. Walking velocity was linearly increased according to the increase of the amplitude parameter A_s. This is one of the good properties of the proposed walking controller because we can easily select the amplitude parameter to achieve the desired walking velocity.

On the other hand, the walking period did not show monotonic change according to the increase of the amplitude A_s as in Figure 7.9b. Walking velocity can be increased by either an increase in walking step or a decrease in walking period (increased walking frequency). We can see that the walking controller used different strategies to increase walking speed with different amplitude A_s.

To evaluate the proposed approach in the real environment, we applied our proposed walking method to the humanoid robot CBi. The parameters for the natural frequency of the controller $\omega_c = 3.14$ rad/s, the coupling constant $K_c = 9.4$,

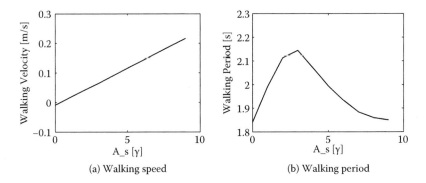

(a) Walking speed

(b) Walking period

FIGURE 7.9 Comparison with different amplitude parameters A_s by setting $A_{hip}^{walk} = A_s$ and $A_{ankle}^{walk} = A_s/2$ in (7.9). The biped model falls over with the parameter $A_s > 9.0°$. (a) Walking speed; (b) walking period.

$A_{hip^{roll}} = A_{ankle^{roll}} = 3.0°$, and $A_{pitch} = 3.5°$ are used. We empirically figured out the proper amplitude parameters $A_{hip}^{walk} = 4.0°$ and $A_{ankle}^{walk} = 2.0°$ for the walking task.

Figure 7.10 shows the successful walking pattern of our humanoid robot. Our proposed method was able to generate successful walking patterns even in the real environment. We showed that the robot was able to walk by only using the simple sinusoidal trajectory, which was composed of at most only two sinusoidal basis functions at each joint, modulated by the detected phase from the center of pressure.

(a) Walking movement

(b) Turning movement

FIGURE 7.10 (a) Successful walking pattern of human-sized humanoid robot CBi. (b) Turning movement.

7.3   UPPER-LEVEL LEARNING SYSTEM

In the upper-level system of the proposed framework, we consider using a mental-simulation–based RL method to adapt the lower-level CPG system to new environments. In particular, we use the approximated Poincaré map model as the mentally simulated environment. To represent the Poincaré map, we use a nonparametric approximation method such as a Gaussian process [38].

A Gaussian process model allows us to estimate the probability distribution of a target nonlinear function with a given covariance. Gaussian processes can estimate the accuracy of the approximated function based on the density of the sampled data. This is beneficial, as it is difficult to uniformly collect data from a real robot due to unknown dynamics. By using this stochastic model RL can take the accuracy of the approximated model into account throughout the learning process.

Algorithm 1

1. Initialize policy parameters.
2. Apply the current policy to the actual robot dynamics and sample data at the defined Poincaré section.
3. Generate a Gaussian process model that represents the stepping and walking dynamics in (7.14).
4. Update policy parameters by applying a reinforcement learning method to the acquired Gaussian process.
5. If the policy is not improved, terminate the iteration.

Otherwise, go back to step 2.

Reinforcement learning (RL), which does not require a precise environmental model, can be a useful technique to improve task performance of real robots. However, one drawback of using RL is that it usually requires a large number of iterations to improve policies. For this reason applications of RL to real environments have been limited [5,11,13,25,31,33,34,37,46].

Methods to improve policy parameters by using inaccurate models have been studied in References [1–3]. In our approach, we propose the use of a task-specific stochastic model instead of using a robot in a real environment to improve a policy. We directly approximate Poincaré maps of stepping and walking dynamics without explicitly identifying rigid-body inertial parameters and without the use of a ground contact model.

In this approach, (1) we modulate the phase of the sinusoids by using the lower-level CPG, and (2) we modulate the amplitude of the sinusoids according to the current state of the inverted pendulum to improve locomotive performance.

In Section 7.3.1, our learning method using approximated Poincaré maps of stepping and walking dynamics is introduced. In Section 7.3.2, we explain how we apply a Gaussian process model to approximate stepping and walking dynamics. In Section 7.3.3, we describe implementation of a RL method for our learning framework, that uses the dynamics approximated by a Gaussian process.

7.3.1 LEARNING FRAMEWORK

For example, we consider the dynamics $\dot{\boldsymbol{\xi}} = \mathbf{g}(\boldsymbol{\xi})$ of a state vector $\boldsymbol{\xi} \in \mathbf{R}^n$. The Poincaré map is a mapping from an $n-1$ dimensional surface S defined in the state space to itself [40]. If $\boldsymbol{\xi}(k) \in S$ is the kth intersection, then the Poincaré map $\mathbf{h}$ is defined by $\boldsymbol{\xi}(k+1) = \mathbf{h}(\boldsymbol{\xi}(k))$. In our study, we defined the section for which the roll derivative of the pendulum equals zero ($\dot{\psi}^{roll} = 0$) (see Figure 7.4).

The policy of the learning system outputs the next action only at this section. We also assume that we can represent the Poincaré map by a stochastic model. If $\mathbf{x}(k)$ is the kth intersection and $\mathbf{u}(k)$ is the control output at the intersection, the model is defined by:

$$\mathbf{x}(k+1) = \mathbf{f}(\mathbf{x}(k), \mathbf{u}(k)) + \mathbf{n}(k) \tag{7.14}$$

where $\mathbf{x} = (\psi^{roll})$ for stepping and $\mathbf{x} = (\psi^{roll}, \psi^{pitch}, \dot{\psi}^{pitch})$ for walking (see Figure 7.4). $\mathbf{n}(k)$ is the noise input. $\mathbf{f}(\mathbf{x}(k), \mathbf{u}(k))$ represents the deterministic part of the Poincaré map.

To improve task performance, we stochastically modulate the amplitude of the sinusoidal patterns according to the current policy $\pi_{\mathbf{w}}$:

$$\pi_{\mathbf{w}}(\mathbf{x}(k), \mathbf{u}(k)) = p(\mathbf{u}(k)|\mathbf{x}(k); \mathbf{w}) \tag{7.15}$$

where $\mathbf{w}$ is the parameter vector of the policy $\pi_{\mathbf{w}}$. In the following sections, we explain how we approximate the stochastic maps (7.14), and how we acquire the control policy $\pi_{\mathbf{w}}$.

In our learning framework, we improve the approximated Poincaré map and the policy iteratively (see Algorithm 1). We first sample data from a simulated model or a real robot by using the current policy for a Gaussian process regression, and then improve the policy by using the Poincaré map approximated by the Gaussian process as in Figure 7.11.

7.3.2 NONPARAMETRIC SYSTEM IDENTIFICATION ON POINCARÉ MAP

We use a Gaussian process (GP) [49] to approximate the Poincaré map in (7.14). Gaussian processes provide us a stochastic representation of an approximated function. With Gaussian processes for regression, we assume that the output values $y_i(i = 1, \ldots, N)$ are sampled from a zero-mean Gaussian whose covariance matrix is a function of the input vectors $\mathbf{z}_i$ ($i = 1, \ldots, N$):

$$p(y_1, \ldots, y_N | \mathbf{z}_1, \ldots, \mathbf{z}_N) = \mathcal{N}(y_1, \ldots, y_N | 0, \mathbf{K}) \tag{7.16}$$

where $\mathbf{K}$ is an covariance matrix of input vectors with elements $\mathbf{K}_{ij} = \kappa(\mathbf{z}_i, \mathbf{z}_j)$. Here, we used a squared exponential covariance function [38]:

$$\kappa(\mathbf{z}_i, \mathbf{z}_j) = v_0 \exp\left(-\sum_{l=1}^{N_d} v_1^l \frac{1}{2} \left(z_i^l - z_j^l\right)^2\right) + v_2 \delta_{ij} \tag{7.17}$$

where δ_{ij} is the Kronecker delta; v_0, v_1^l, and v_2 are parameters for the covariance matrix; N_d denotes the number of input dimensions. These parameters can be optimized by using a type-II maximum likelihood method [49]. A covariance function

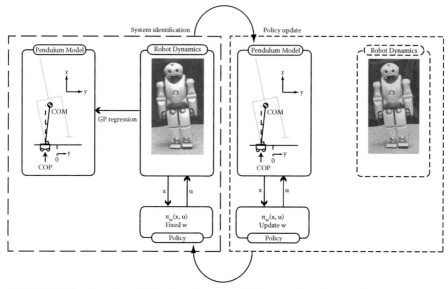

FIGURE 7.11   Steps 2 and 3 in Algorithm 1. (Left) System identification. Apply the current policy with fixed policy parameters **w** to the actual robot dynamics and sample data. The sampled data are used to construct the simulated inverted pendulum model. (Right) Policy update. Update the policy parameters by interacting with the simulated inverted pendulum model.

introduces similarity between datapoints. By using the squared exponential covariance function, closely placed datapoints are similar. This similarity measure works well in many applications and is widely used [6,38].

The conditional distribution of an output y_{N+1} corresponding to a new input $\mathbf{z}_{N+1}$ is given as

$$p(y_{N+1}|\mathbf{z}_1, \ldots, \mathbf{z}_N, \mathbf{z}_{N+1}, y_1, \ldots, y_N) = \mathcal{N}(y_{N+1}|\mu, \sigma^2) \tag{7.18}$$

where

$$\mu = \mathbf{k}(\mathbf{z}_{N+1})^\top \mathbf{K}^{-1}\mathbf{y} \tag{7.19}$$

$$\sigma^2 = \kappa(\mathbf{z}_{N+1}, \mathbf{z}_{N+1}) - \mathbf{k}(\mathbf{z}_{N+1})^\top \mathbf{K}^{-1}\mathbf{k}(\mathbf{z}_{N+1}) \tag{7.20}$$

Here the vectors $\mathbf{y}$ and $\mathbf{k}(\mathbf{z}_{N+1})$ are defined as $\mathbf{y} = (y_1, \ldots, y_N)$ and $\mathbf{k}(\mathbf{z}_{N+1}) = [\kappa(\mathbf{z}_1, \mathbf{z}_{N+1}), \ldots, \kappa(\mathbf{z}_N, \mathbf{z}_{N+1})]^\top$, respectively.

We use the GP model to represent how the pendulum model in Figure 7.4 behaves with a controller, a robot, actuators, and ground contact models.

The input vector $\mathbf{z}$ for the Gaussian process is composed of the current state $\mathbf{x}(k)$ and control input $\mathbf{u}(k)$: $\mathbf{z} = (\mathbf{x}(k)^\top, \mathbf{u}(k)^\top)^\top$. The output value y is a component of the state vector at the next intersection $\mathbf{x}(k+1)$, and the control output $\mathbf{u}(k)$ is used to modulate parameters of the stepping and walking controllers. In this study, we modulate amplitudes of the sinusoidal patterns that are used as desired joint angles.

The predictive distribution represented by this Gaussian process method is used to improve biped stepping and walking controllers.

7.3.3 POLICY IMPROVEMENT BY USING A REINFORCEMENT LEARNING METHOD

Here, we explain how we applied RL to our biped stepping and walking tasks. We used a policy gradient method proposed by Kimura and Kobayashi [23] to implement the RL framework. This learning method was used in biped learning studies [25,46].

Policy gradient methods are considered as robust learning methods when states for the learning system are partially observable [4,21,27]. Convergence properties of policy gradient methods with function approximators have been studied in References [24,43].

The basic goal is to find a policy $\pi_{\mathbf{w}}(\mathbf{x}, \mathbf{u}) = p(\mathbf{u}|\mathbf{x}; \mathbf{w})$ that maximizes the expectation of the discounted accumulated reward:

$$E\{V(k)|\pi_{\mathbf{w}}\} = E\left\{\sum_{i=k}^{\infty} \gamma^{i-k} r(i) \middle| \pi_{\mathbf{w}}\right\} \tag{7.21}$$

where r denotes reward, $V(k)$ is the actual return, $\mathbf{w}$ is the parameter vector of the policy $\pi_{\mathbf{w}}$, and γ, $0 \le \gamma < 1$, is the discount factor.

In policy gradient methods, we calculate the gradient direction of the expectation of the actual return with respect to the parameters of a policy $\mathbf{w}$. We can estimate the expectation of the gradient direction as suggested in Reference [23]:

$$\frac{\partial}{\partial \mathbf{w}} E\{V(0)|\pi_{\mathbf{w}}\} \approx E\left\{\sum_{k=0}^{\infty}(V(k) - \hat{V}(\mathbf{x}))\frac{\partial \ln \pi_{\mathbf{w}}}{\partial \mathbf{w}} \middle| \pi_{\mathbf{w}}\right\} \tag{7.22}$$

where $\hat{V}(\mathbf{x})$ is an approximation of the value function for a policy $\pi_{\mathbf{w}}$: $V^{\pi_{\mathbf{w}}}(\mathbf{x}) = E\{V(k)|\mathbf{x}(k) = \mathbf{x}, \pi_{\mathbf{w}}\}$.

7.3.3.1 Value Function Approximation

The value function is approximated using a normalized Gaussian network [10]:

$$\hat{V}(\mathbf{x}) = \sum_{i=1}^{N} w_i^v b_i(\mathbf{x}) \tag{7.23}$$

where w_i^v is an ith parameter of the approximated value function, and N is the number of basis functions $b_i(\mathbf{x})$. An approximation error of the value function is represented by the temporal difference (TD) error [42]:

$$\delta(k) = r(k+1) + \gamma \hat{V}(\mathbf{x}(k+1)) - \hat{V}(\mathbf{x}(k)) \tag{7.24}$$

We update the parameters of the value function approximator using the TD(0) method [42]:

$$w_i^v(k+1) = w_i^v(k) + \alpha\delta(k)b_i(\mathbf{x}(k)) \tag{7.25}$$

where α is the learning rate.

7.3.3.2 Policy Parameter Update

We update the parameters of a policy $\mathbf{w}$ by using the estimated gradient direction in (7.22). We can estimate the gradient direction by using the TD error as described in Reference [23]:

$$E\left\{\sum_{k=0}^{\infty}(V(k) - \hat{V}(\mathbf{x}(k)))\frac{\partial \ln \pi_{\mathbf{w}}}{\partial \mathbf{w}}\bigg|\pi_{\mathbf{w}}\right\} = E\left\{\sum_{k=0}^{\infty}\delta(k)e(k)\bigg|\pi_{\mathbf{w}}\right\} \qquad (7.26)$$

where $\mathbf{e}$ is the eligibility trace of the parameter $\mathbf{w}$. Eligibility traces indicate which parameters are eligible for the TD error and are updated as

$$\mathbf{e}(k+1) = \eta\mathbf{e}(k) + \frac{\partial \ln \pi_{\mathbf{w}}(\mathbf{x}(k), \mathbf{u}(k))}{\partial \mathbf{w}}\bigg|_{\mathbf{w}=\mathbf{w}(k)} \qquad (7.27)$$

where η is the decay factor for the eligibility trace. Equation (7.26) can be derived if the condition $\eta = \gamma$ is satisfied. The parameter $\mathbf{w}$ is updated as

$$\mathbf{w}(k+1) = \mathbf{w}(k) + \beta\delta(k)\mathbf{e}(k) \qquad (7.28)$$

7.3.3.3 Biped Stepping and Walking Policies

We construct the biped stepping and walking policies based on a normal distribution:

$$\pi_{\mathbf{w}}(\mathbf{x}, \mathbf{u}) = \mathcal{N}(\mu(\mathbf{x}; \mathbf{w}^\mu), \Sigma(\mathbf{x}; \mathbf{w}^\sigma)) \qquad (7.29)$$

where $\mathbf{u}$ is the output vector and Σ is the covariance matrix of the policy $\pi_{\mathbf{w}}$. In this study, we defined the covariance matrix as a diagonal matrix, where the jth diagonal element is represented as σ_j. The jth element of the mean output μ is modeled by a normalized Gaussian network:

$$\mu_j(\mathbf{x}) = \sum_{i=1}^{N} w_i^{\mu_j} b_i(\mathbf{x}) \qquad (7.30)$$

Here, $w_i^{\mu_j}$ denotes the ith parameter for the jth output of the policy $\pi_{\mathbf{w}}$, and N is the number of basis functions. We represent the diagonal element of the covariance matrix Σ using a sigmoid function [23]:

$$\sigma_j(\mathbf{x}) = \frac{\sigma_0}{1 + \exp(-\sigma_j^w(\mathbf{x}))}, \quad \text{where } \sigma_j^w(\mathbf{x}) = \sum_{i=1}^{N} w_i^{\sigma_j} b_i(\mathbf{x}) \qquad (7.31)$$

and σ_0 denotes the scaling parameter. $w_i^{\sigma_j}$ denotes the ith parameter for the jth diagonal element of the covariance matrix. We update the parameters by applying the update rules in (7.28) and (7.27).

We use same basis functions $b_i(\mathbf{x})$ in (7.23), (7.30), and (7.31).

TABLE 7.1
Parameters of the Periodic Pattern Generator for Each Experiment

	Simulated Model		Real Robot	
	Stepping	Walking	Stepping	Walking
ω_c	3.5	3.5	6.3	6.3
K_c	10.0	10.0	9.4	9.4
$A_{hip^{roll}}$	4.0	3.5	2.5	5.0
$A_{ankle^{roll}}$	4.0	3.5	8.5	9.0
A_{pitch}	6.0	7.0	4.0	4.0

7.3.4 SIMULATION

We applied our proposed method to the simple 3D simulated biped model in Figure 7.2a. Parameters used in the controllers are summarized in Table 7.1. Parameters used in the policy gradient method are summarized in Table 7.2.

7.3.4.1 Improvement of Biped Stepping Performance

We applied our proposed method to improve stepping in place. We selected amplitudes of the pitch joint movements as the control output: $\mathbf{u}(k) = A^{step}(k)$. Then, the desired joint angles in (7.8) become

$$\theta^d_{hip^{pitch}}(\phi_c) = (A_{pitch} + \underline{A^{step}})\sin(\phi_c) + \theta^{res}_{hip^{pitch}}$$
$$\theta^d_{knee^{pitch}}(\phi_c) = -2(A_{pitch} + \underline{A^{step}})\sin(\phi_c) + \theta^{res}_{knee^{pitch}}$$
$$\theta^d_{ankle^{pitch}}(\phi_c) = -(A_{pitch} + \underline{A^{step}})\sin(\phi_c) + \theta^{res}_{ankle^{pitch}} \qquad (7.32)$$

We flip the sign of the roll angle ψ^{roll} in the state vector $\mathbf{x}$ when the COP sign in the lateral (y) direction (see Figure 7.4a) changes so that we can use the same policy for the left stance phase and the right stance phase.

We defined the target of the stepping task to keep the desired state at ψ^{roll}_d. We use a reward function:

$$r = -K(\psi^{roll}_d - \psi^{roll})^2 \qquad (7.33)$$

for this stepping task, where the desired roll angle $\psi^{roll}_d = 2.0°$ and $K = 0.1$. The learning system also receives a negative reward $r = -1$ if the biped model falls over. We consider that the biped model falls over when the condition $\psi^{roll} < 0.0°$ is satisfied (the COM is outside the stance foot).

TABLE 7.2
Parameters of Policy Gradient Method

γ	α	η	β	σ_0
0.95	0.3	0.3	0.3	0.5

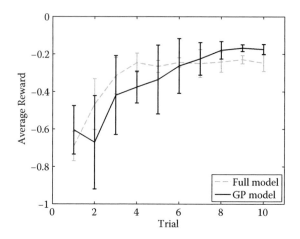

FIGURE 7.12   A comparison of learning performance. At each visit to the Poincaré section, a reward is given according to the performance of a policy. The average reward (vertical axis) shows the averaged acquired reward at each visit to the Poincaré section. The solid line represents the performance of the proposed learning method that used the acquired Gaussian process model. The dashed line represents the learning performance of the learning method that used the full dynamics simulation. The result shows the average performance of five simulation runs. Error bars show the standard deviation.

For the Gaussian process model, we sampled 20 datasets $\{\mathbf{x}(k+1), \mathbf{x}(k), \mathbf{u}(k)\}$ from the simulated environments (at step 2 of Algorithm 1).

We compare learning performance of the proposed method using the Gaussian process model with that using the full model. The full model means that we run the dynamics simulator of the biped model to learn stepping policies.

Figure 7.12 compared learning performance on the stepping task by using the GP model and by using the full model. At each visit to the Poincaré section, a reward is given according to the performance of a policy. The average reward (vertical axis) shows the average acquired reward at each visit to the Poincaré section.

Although the variance of the learning performance is larger in the early stage of the learning process if we use the GP model, the performance of the acquired policies through the GP model is comparable to the policies acquired through the full model after 10 learning trials. Thus, at least for this particular example, this demonstrates that a full dynamic model is not necessary in order to acquire a policy to achieve sufficient task performance. Even in extensive simulation studies we can save computation time, and still may be able to acquire comparable performance.

Furthermore, the acquired stepping policies based on the GP model slightly outperformed the policies acquired by the full model after the eighth trial. This difference was significant in terms of the Student's t-test with significance level 0.05. The gradient estimation of policy parameters can be biased with limited sample data, thus using the approximated model with a Bayesian estimation method such as the Gaussian process regression would generate better sample data for the gradient estimation. Therefore, it would be interesting to compare Bayesian RL approaches [8,16] with the proposed method as a future study in addition to applying the

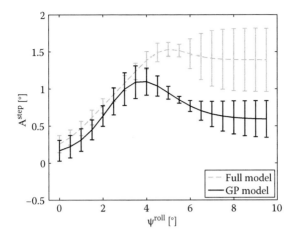

FIGURE 7.13 Acquired stepping policies. The solid line represents acquired policies using the GP model. The dashed line represents acquired policies using the full model. The results show an average policy of five simulation runs. Error bars show the standard deviation.

proposed method to other examples to show how this fact can be generalized to other applications.

Figure 7.13 shows policies acquired by using the GP model and the full model. Similar policies were acquired in the region $0.0° < \psi^{roll} < 4.0°$, the region of the state space where the policies are mainly explored. Outside the region $4.0° \leq \psi^{roll}$, the policies are different because the data are not frequently sampled from this region. From 3 to 5 basis functions are allocated to represent the mean output of the policies in Equation (7.24). Therefore, $3\sim 5$ parameters were needed to be learned for the mean output. Because we adaptively allocated the basis functions, the number of basis functions is different in different simulation runs [31].

We also investigated the robustness of an acquired stepping controller. Figure 7.14 shows the return map of the pendulum state ψ^{roll}. Because we change the output $\mathbf{u}(k)$ only at the Poincaré section, it takes two steps to evaluate the control performance of the control output from an acquired policy. We showed the relationship between $\psi^{roll}(k)$ and $\psi^{roll}(k + 2)$. To generate the return map, we disturbed the stepping controller for 0.2 s right after the COP crosses zero by pushing the biped model in the horizontal (y) direction (see Figure 7.5) with randomly generated force F, where F is sampled from a uniform distribution over $-40 \leq F \leq 40$ N.

The dashed line shows the linear approximation of the return map. Because the coefficient of the linear approximation is much smaller than one, this result shows the robustness of the acquired stepping policy. The approximated fixed point $\psi^{roll} = 2.3°$ is indicated by the circle and is close to the desired $\psi_d^{roll} = 2.0°$.

7.3.4.2 Improvement of Biped Walking Performance

We also apply our proposed method to improve walking performance. We modulate the amplitude A_{hip}^{walk} and A_{ankle}^{walk} in (7.9) to generate forward movement for the biped walking task. The target of the walking task is to increase the walking velocity. The

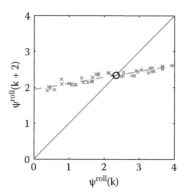

FIGURE 7.14 Return map of ψ^{roll}. We show the relationship between $\psi^{roll}(k)$ and $\psi^{roll}(k+2)$. The dashed line shows the linear approximation of the return map.

angular velocity of the pendulum $\dot\psi^{pitch}$ (see Figure 7.4c) at the Poincaré section corresponds to the walking velocity, therefore we use the reward function:

$$r = K(\dot\psi^{pitch}) \tag{7.34}$$

for this biped walking task, where $K = 0.1$. The learning system also receives a negative reward $r = -1$ if the biped model falls over: $|\psi^{roll}| > 15.0°$ or $|\psi^{pitch}| > 15.0°$.

For the Gaussian process model, we sampled 100 datasets from the simulated environments. Figure 7.15 shows the initial performance of the biped walking policy. Figure 7.15b shows the walking performance after one iteration of the proposed method. One iteration means (1) acquisition of an approximated Poincaré map from sampled data; (2) acquisition of a stepping policy using the acquired Poincaré map; and (3) application of the acquired policy to the simulated robot as in Figure 7.11.

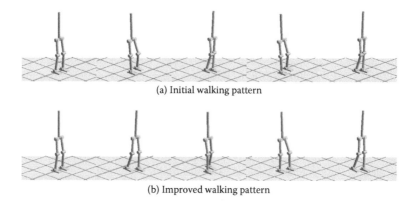

(a) Initial walking pattern

(b) Improved walking pattern

FIGURE 7.15 (a) Initial walking pattern. The thick line represents the starting position. Initially, the simulated robot explores around the starting position. Time proceeds from left to right. (b) Improved walking pattern after one iteration of the proposed learning process. Walking speed is 0.14 m/s.

This result suggests that approximated dynamics can be used to improve biped walking performance.

7.3.5 Experimental Result

We applied our proposed method to a small-size humanoid robot (see Figure 7.2c). Parameters used in the controllers are summarized in Table 7.1.

7.3.5.1 Improvement of Biped Stepping Performance

We use the roll angle $\mathbf{x} = (\psi^{roll})$ defined by the pendulum model (see Figure 7.4) as the state. We select amplitudes of the ankle roll movements as the control output: $\mathbf{u}(k) = A_{ankle}^{step}$. Then, the desired joint angle in (7.7) becomes:

$$\theta_{ankle^{roll}}^{d}(\phi_c) = -(A_{ankle^{roll}} + A_{ankle}^{step})\sin(\phi_c) \tag{7.35}$$

Because the real robot has relatively wide feet, modulating the ankle roll joints is an easy way to control the roll angle $\mathbf{x} = (\psi^{roll})$. Automatic selection of the proper action variables for a given task would be an interesting research topic for future study.

We use the reward function:

$$r = -K(\psi_d^{roll} - \psi^{roll})^2 \tag{7.36}$$

for this stepping task, where the desired roll angle $\psi_d^{roll} = 0.0°$ and $K = 0.1$. The learning system also receives a negative reward $r = -1$ if the biped model falls over: $|\psi^{roll}| > 10.0°$.

For the Gaussian process model, we sampled 20 datasets from the real environment. Figure 7.16 shows an approximated Poincaré map for stepping dynamics of the small-size humanoid robot using a Gaussian process. Here we define the input vector as $\mathbf{z} = (\psi^{roll}(k), A_{ankle}^{step}(k))$ and the output as $y = (\psi^{roll}(k+1))$. We apply the reinforcement

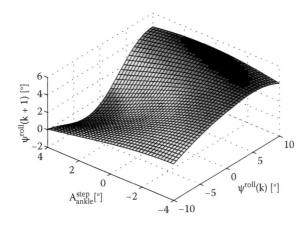

FIGURE 7.16 Approximated stepping dynamics of the small-size humanoid robot by a Gaussian process. The input vector is defined as $\mathbf{z} = (\psi^{roll}(k), A_{ankle}^{step}(k))$, and the output is defined as $y = (\psi^{roll}(k+1))$. (See (7.19).)

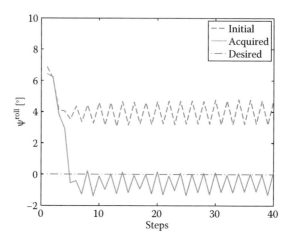

FIGURE 7.17   The roll angle ψ^{roll} at the Poincaré section $\dot{\psi}^{roll} = 0$ (solid line). The dash-dotted line represents the desired angle for this stepping task. We used a policy acquired by the proposed learning method.

learning algorithm to this acquired Poincaré map for stepping dynamics to improve stepping performance.

Figure 7.17 shows the roll angle ψ^{roll} at the Poincaré section $\dot{\psi}^{roll} = 0$. This result suggests that a stepping policy was acquired by using our proposed method, and it can keep the roll state ψ^{roll} around the desired angle (0.0°). An average angle from 10 to 40 steps was 0.12°.

Figure 7.18 shows the acquired stepping movement of the real robot after one iteration of the proposed learning process.

Again, one iteration means (1) acquisition of an approximated Poincaré map from sampled real data as in the left-hand side of Figure 7.11; (2) acquisition of a stepping policy using the acquired Poincaré map as in the right-hand side of Figure 7.11; and (3) application of the acquired policy to the real robot.

7.3.5.2   Improvement of Biped Walking Performance

We also applied our method to improve the biped walking performance of the humanoid robot. We used $\mathbf{x} = (\psi^{roll}, \psi^{pitch}, \dot{\psi}^{pitch})$ as the state and modulated A_{hip}^{walk} and A_{ankle}^{walk} in (7.9) as the action of the learning system.

We used the reward function based on the pitch derivative $\dot{\psi}^{pitch}$ that corresponds to the walking velocity:

$$r = \frac{1}{T} \int_0^T \dot{\psi}^{pitch} dt \tag{7.37}$$

for this walking task, where T is the time required for moving from a Poincaré section to the next section defined in Section 7.3.1, which is approximately the time for one

FIGURE 7.18 Acquired stepping movement of the real robot after one iteration of the proposed learning process. One iteration means (1) acquisition of an approximated Poincaré map from sampled real data; (2) acquisition of a stepping policy using the acquired Poincaré map; and (3) application of the acquired policy to the real robot.

cycle of walking. We used the time integral of the angular velocity $\dot{\psi}^{pitch}$ rather than the instantaneous angular velocity at the defined section because the instantaneous value was not consistently related to the walking velocity in the real environment. This reward cannot be simply represented by the state **x** at the section, therefore we also approximated a stochastic model of the reward function from the data acquired from the real environment by using another Gaussian process. The learning system also received a negative reward $r = -1$ if the biped model fell over: $|\psi^{roll}| > 10.0°$ or $|\psi^{pitch}| > 10.0°$.

For the Gaussian process model, we sampled 100 datasets from the real environments. Figure 7.19a shows the initial performance of the biped walking controller. The initial policy only generates random steps and the humanoid robot could not walk forward.

Figure 7.19b shows the improved walking performance after the first iteration of our learning framework. The humanoid robot could walk forward with the improved policy.

Figure 7.19c shows the improved walking performance after the second iteration of our learning framework. The humanoid robot could walk faster. We also tried additional learning iterations. However, the walking performance did not significantly improve. We may need to use different state representations and/or reward functions to acquire additional improvement of the walking policy.

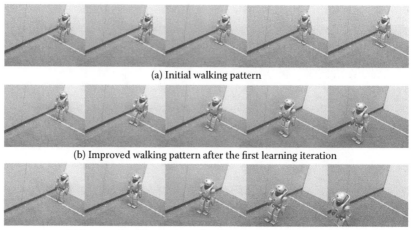

(a) Initial walking pattern

(b) Improved walking pattern after the first learning iteration

(c) Improved walking pattern after the second learning iteration

FIGURE 7.19 Initial walking pattern. The line represents the starting position. Initially, the humanoid robot explores around the starting position. Time proceeds from left to right.

7.4 CONCLUSION

In this chapter, a two-layered biologically inspired biped learning model was introduced. In the lower-level CPG, the center of pressure position and velocity were used to detect the phase of the lateral robot dynamics. Evidence in biological locomotion studies supports our work in part [26,50]. The detected phase of the robot dynamics was used to modulate sinusoidal joint trajectories. The modulated trajectories enabled a human-sized humanoid robot CBi to generate successful stepping and walking patterns. Because the angular frequency in Equation (7.5) is continuously changing during stepping and walking, not only the frequency of the controller changes toward the resonant frequency and excites the robot dynamics but also the time course of the sinusoidal patterns is modulated.

In the upper-level reinforcement learning module, a nonparametric representation of approximated Poincaré maps was used for biped stepping and walking dynamics and reinforcement learning (RL) to improve task performance. In this approach, we first approximated Poincaré maps for stepping and walking dynamics by using data from a simulated model or a real robot, then used the approximated maps for RL to improve stepping and walking policies. We explored using a Gaussian process to approximate the Poincaré maps. By using a Gaussian process, we can estimate a probability distribution of the target nonlinear functions with a given covariance. We showed that we could improve stepping and walking policies by using a RL method with approximated models both in simulated and real environments. Because we used a different reward for the stepping and walking tasks, the results showed that our learning framework can improve policies for different objective functions. We applied the proposed control approach to a small humanoid robot.

The robustness of the resulting stepping policy was evaluated by pushing the biped model in the horizontal direction and analyzing the return map. We also showed robustness of the resulting walking policy by applying the policy to a real environment.

Compared to other biped learning approaches (e.g., References [5,34,46]), we do not directly use real environments to improve biped walking policies. Therefore, our learning framework requires a smaller number of samples from the real environment and can be applied to robots that have many degrees of freedom. On the other hand, our learning method relies on environmental models. We take the reliability of an acquired model into account in a policy improvement process by representing the model with a probability distribution.

Other reward functions can be adopted for the stepping and the biped walking task. If we can measure energy consumption of real biped robots, we can consider the energy cost as a negative reward (penalty). Similarity between walking patterns of biped robots and human walking patterns could be used to acquire a policy to generate natural-looking walking patterns. Ground reaction force can be used as a negative reward to reduce impact forces when the swing leg contacts the ground.

REFERENCES

1. Abbeel, P., Quigley, M., and Ng, A.Y. Using inaccurate models in reinforcement learning. In *Proceedings of the 23rd international conference on Machine learning*, New York: ACM Press, 2006, pp. 1–8.
2. Atkeson, C.G. Nonparametric model-based reinforcement learning. In M.I. Jordan, M. Kearns, and S. Solla (Eds.), *Advances in Neural Information Processing Systems 10*, Cambridge, MA: MIT Press, 1998, pp. 1008–1014.
3. Atkeson, C.G. and Schaal, S. Robot learning from demonstration. In *Proceedings of the 14th International Conference on Machine Learning*, San Francisco: Morgan Kaufmann, 1997, pp. 12–20.
4. Baird, L.C. and Moore, A.W. Gradient descent for general reinforcement learning. In *Advances in Neural Information Processing Systems 11*, Cambridge, MA: MIT Press, 1999.
5. Benbrahim, H. and Franklin, J. Biped dynamic walking using reinforcement learning. *Robot. Auton. Syst.* **22**:283–302, 1997.
6. Bishop, C.M. *Pattern Recognition and Machine Learning*. New York: Springer, 2006.
7. Cheng, G. Hyon, S.-H., Morimoto, J., Ude, A., Hale, J.G., Colvin, G., Scroggin, W. and Jacobsen, S.C. CB: A Humanoid Research Platform for Exploring Neuroscience. *Adv. Robot.* **21**(10):1097–1114, 2007.
8. Dearden, R., Friedman, N., and Andre, D. Model based Bayesian exploration. In *Proceedings of Fifteenth Conference on Uncertainty in Artificial Intelligence*, San Francisco: Morgan Kaufmann, pp. 457–464, 1999.
9. Doi, M., Hsegawa, Y., and Fukuda, T. Passive trajectory control of the lateral motion in bipedal walking. In *IEEE International Conference on Robotics and Automation*, New Orleans, pp. 3049–3054, 2004.
10. Doya, K. Reinforcement learning in continuous time and space. *Neural Comput.* **12**(1):219–245, 2000.
11. Endo, G., Morimoto, J., Matsubara, T., Nakanishi, J., and Cheng, G. Learning CPG-based biped locomotion with a policy gradient method: Application to a humanoid robot. *Int. J. Robot. Res.* **27**(2):213–228, 2008.
12. Endo, G., Morimoto, J., Nakanishi, J., and Cheng, G. An empirical exploration of a neural oscillator for biped locomotion control. In *Proceedings of IEEE 2004 International Conference on Robotics and Automation*, New Orleans, pp. 3063–3042, 2004.

13. Endo, G., Morimoto, J., Nakanishi, J., Matsubara, T., and Cheng, G. Learning CPG sensory feedback with policy gradient for biped locomotion for a full-body humanoid. In *The Twentieth National Conference on Artificial Intelligence*, pp. 1267–1273, 2005.

14. Endo, G., Nakanishi, J., Morimoto, J., and Cheng, G. Experimental Studies of a Neural Oscillator for Biped Locomotion with QRIO. In *IEEE International Conference on Robotics and Automation*, Barcelona, pp. 598–603, 2005.

15. Fukuoka, Y. and Kimura, H. Adaptive dynamic walking of a quadruped robot on irregular terrain based on biological concepts. *Intl. J. Robot. Res.* **22**(2):187–202, 2003.

16. Ghavamzadeh, M. and Engel, Y. Bayesian Policy Gradient Algorithms. In B. Scholkopf, J. Platt, and T. Hofmann, (Eds.) *Advances in Neural Information Processing Systems 19*, Cambridge, MA: MIT Press, pp. 457–464, 2007.

17. Grillner, S. Locomotion in vertebrates: Central mechanisms and reflex interaction. *Physiol. Rev.* **55**:367–371, 1975.

18. Grillner, S. Neurobiological bases of rhythmic motor acts in vertebrates. *Science*, **228**:143–149, 1985.

19. Hirai, K., Hirose, M., and Takenaka, T. The development of honda humanoid robot. In *Proceedings of the 1998 IEEE International Conference on Robotics and Automation*, pp. 160–165, 1998.

20. Hyon, S. and Emura, T. Symmetric walking control: Invariance and global stability. In *IEEE International Conference on Robotics and Automation*, Barcelona, pp. 1456–1462, 2005.

21. Jaakkola, T., Singh, S.P., and Jordan, M.I. Reinforcement learning algorithm for partially observable markov decision problems. In G. Tesauro, D. Touretzky, and T. Leen (Eds.), *Advances in Neural Information Processing Systems*, Vol. 7, Cambridge, MA: MIT Press, pp. 345–352, 1995.

22. Kajita, S., Kanehiro, F., Kaneko, K., Fujiwara, K., Yokoi, K., and Hirukawa, H. Biped walking pattern generation by a simple three-dimensional inverted pendulum model. *Adv. Robot.* **17**(2):131–147, 2004.

23. Kimura, H. and Kobayashi, S. An analysis of actor/critic algorithms using eligibility traces: Reinforcement learning with imperfect value functions. In *Proceedings of the 15th International Conference on Machine Learning*, pp. 284–292, 1998.

24. Konda, V.R. and Tsitsiklis, J.N. Actor-critic algorithms. *SIAM J. Control Optim.* **42**(4):1143–1166, 2003.

25. Matsubara, T., Morimoto, J., Nakanishi, J., Sato, M., and Doya, K. Learning CPG-based biped locomotion with a policy gradient method. *Robot. Auton. Syst.* **54**(11):911–920, 2006.

26. McMahon, T.A. *Muscles, Reflexes, and Locomotion*. Princeton, NJ: Princeton University Press, 1984.

27. Meuleau, N., Kim, K.E., and Kaelbling, L.P. Exploration in gradient-based reinforcement learning. *Tech. rep., AI Memo 2001-003*, Cambridge, MA: MIT, 2001.

28. Miura, H. and Shimoyama, I. Dynamical walk of biped locomotion. *Intl. J. Robot. Res.* **3**(2):60–74, 1984.

29. Miyakoshi, S., Taga, G., Kuniyoshi, Y., and Nagakubo, A. Three dimensional bipedal stepping motion using neural oscillators - Towards humanoid motion in the real world. In *IEEE/RSJ International Conference on Intelligent Robots and Systems*, Vol. 1, Victoria, Canada, pp. 84–89, 1998.

30. Morimoto, J. and Atkeson, C.G. Nonparametric representation of an approximated Poincaré map for learning biped locomotion. *Auton. Robot.* **27**(2):131–144, 2009.

31. Morimoto, J. and Doya, K. Acquisition of stand-up behavior by a real robot using hierarchical reinforcement learning. *Robot. Auton. Syst.* **36**:37–51, 2001.

32. Morimoto, J., Endo, G., Nakanishi, J., Hyon, S., Cheng, G., Atkeson, C.G., and Bentivegna, D. Modulation of simple sinusoidal patterns by a coupled oscillator model for biped walking. In *Proceedings of the 2006 IEEE International Conference on Robotics and Automation*, pp. 1579–1584, 2006.

33. Morimoto, J., Nakanishi, J., Endo, G., Cheng, G., Atkeson, C.G., and Zeglin, G. Poincaré-map-based reinforcement learning for biped walking. In *Proceedings of the 2005 IEEE International Conference on Robotics and Automation*, pp. 2392–2397, 2005.

34. Morimoto, J. and Atkeson, C.G. Learning biped locomotion: Application of Poincaré-map-based reinforcement learning. *IEEE Robot. Autom. Mag.* **14**(2):41–51, 2007.

35. Nagasaka, K., Inaba, M., and Inoue, H. Stabilization of dynamic walk on a humanoid using torso position compliance control. In *Proceedings of 17th Annual Conference on Robotics Society of Japan*, pp. 1193–1194, 1999.

36. Nakanishi, J., Morimoto, J., Endo, G., Cheng, G., Schaal, S., and Kawato, M. Learning from demonstration and adaptation of biped locomotion. *Robot. Auton. Syst.* **47**:79–91, 2004.

37. Peters, J. and Schaal, S. Reinforcement learning of motor skills with policy gradients. *Neural Netw.* **21**(4):682–697, 2008.

38. Rasmussen, C.E. and Williams, C.K.I. *Gaussian Processes for Machine Learning.* Cambridge, MA: MIT Press, 2006.

39. Shik, M.L. and Orlovsky, G.N. Neurophysisology of locomotion automatism. *Physiol. Rev.* **56**:465–501, 1976.

40. Strogatz, S.H. *Nonlinear Dynamics and Chaos.* Reading, MA: Addison-Wesley, 1994.

41. Sugihara, T. and Nakamura, Y. Whole-body cooperative COG control through ZMP manipulation for humanoid robots. In *IEEE International Conference on Robotics and Automation*, Washington DC, 2002.

42. Sutton, R.S. and Barto, A.G. *Reinforcement Learning: An Introduction.* Cambridge, MA: MIT Press, 1998.

43. Sutton, R.S., McAllester, D., Singh, S., and Mansour, Y. Policy gradient methods for reinforcement learning with function approximation. In *Advances in Neural Information Processing Systems 12*, Cambridge, MA: MIT Press, pp. 1057–1063, 2000.

44. Sutton, R.S. Planning by incremental dynamic programming. In *Proceedings of the Eighth International Conference on Machine Learning*, San Francisco: Morgan Kaufmann, pp. 353–357, 1991.

45. Taga, G., Yamaguchi, Y., and Shimizu, H. Self-organized control in bipedal locomotion by neural oscillators in unpredictable environment. *Biol. Cybern.* **65**:147–159, 1991.

46. Tedrake, R., Zhang, T.W., and Seung, H.S. Stochastic policy gradient reinforcement learning on a simple 3D biped. In *Proceedings of the 2004 IEEE/RSJ International Conference on Intelligent Robots and Systems*, pp. 2849–2854, 2004.

47. Tsuchiya, K., Aoi, S., and Tsujita, K. Locomotion control of a biped locomotion robot using nonlinear oscillators. In *Proceedings of the IEEE/RSJ International Conference on Intelligent Robots and Systems*, Las Vegas, pp. 1745–1750, 2003.

48. Westervelt, E.R., Buche, G., and Grizzle, J.W. Experimental validation of a framework for the design of controllers that induce stable walking in planar bipeds. *Int. J. Robot. Res.* **23**(6):559–582, 2004.

49. Williams, C.K.I. and Rasmussen, C.E. Gaussian processses for regression. In *Advances in Neural Information Processing Systems*, Vol. 8, Cambridge, MA: MIT Press, pp. 514–520, 1996.

50. Winter, D.A. *Biomechanics and Motor Control of Human Movement*, 2nd edition, New York: John Wiley & Sons, 1990.

8 Humanoid Motor Control: Dynamics and the Brain

Sang-Ho Hyon

CONTENTS

This chapter describes how we make humanoid robots robustly perform skillful full-body motor tasks under complex environments and limited resources. Starting from the foundation of full-body humanoid dynamics, we introduce dynamics-based controllers, then discuss how to make the algorithms realistic and practical. As the key motor function of humanoids, we mainly focus on dynamic bipedal balancing. Basically we utilize wide knowledge from naive robotics and control theory, then introduce some neuroscientific clues to improve the control performance on real robots. This is not at all a theory-oriented study, but a realistic experimental approach. Sufficient experimental results on a life-size hydraulic humanoid robot are attached.

8.1 BASIC COMPONENTS OF FULL-BODY HUMANOID DYNAMICS

As long as physics is involved in the motion of a body, it is natural to assume animals are utilizing it when controlling the motion. Therefore, it is assumed that a highly sophisticated model-based controller is running in animal central nervous systems (CNS). The model is not necessarily the dynamics of a multilink rigid body only, because complex neuromusculo dynamics is involved in the low-level nervous system. Nevertheless, the multilink rigid-body dynamics would be the best starting point to investigate what kind of controller must be embedded in CNS and other controller levels. In contrast to fixed-base robot dynamics, however, we must think about the dynamics of free-flying robots with constraints, whose mathematical formulation is described below.

8.1.1 EQUATION OF MOTION

Consider a multi-DoF humanoid robot contact with the ground, as shown in Figure 8.1. Let $r \in \mathcal{R}^3$ in Σ_W be some translational position coordinate (e.g., base position), $\phi \in \mathcal{R}^3$ be the orientation (Euler angle) of the base, and $\theta \in \mathcal{R}^n$ be the joint angles. For simplicity we do not show the formal SE formulation or quarternian description because the latter discussion is not affected by this. Using the generalized coordinates $q = [r, \phi, \theta]^T \in \mathcal{R}^{6+n}$ the exact nonlinear dynamics of humanoids with the constraint can be derived using a standard Lagrangian formulation with constraints [1]:

$$I(q)\ddot{q} + C(q, \dot{q}) + G(q) = u + E(q)^T \lambda \tag{8.1}$$

$$E(q)\dot{q} = 0 \tag{8.2}$$

where $I(q) \in \mathcal{R}^{(6+n) \times (6+n)}$ is the inertia matrix; $C(q, \dot{q}) \in \mathcal{R}^{6+n}$, respectively, represent the centrifugal and Coriolis terms; $G(q) \in \mathcal{R}^{6+n}$ is the gravity term; $u = [0, \tau]^T \in \mathcal{R}^{6+n}$ is the generalized force; f_p is the constraint force; and $E(q) \in \mathcal{R}^{\alpha \times n}$ is the associated Jacobian.

The constraints can be removed or added during the task. For example, double support or single support have different sets of constraints, both appear during a biped walking task. Furthermore, one can choose any kind of contact representation, for example, multiple point contacts or a pair of single point and orientation constraints, which results in translational constraint force only, or translational plus rotational force (moment), respectively. Apparently, the physical effect is equivalent. For a multipoint contact problem, there arise ill-posedness problems anyway where some optimization should be applied to determine each contact force. That is the problem.

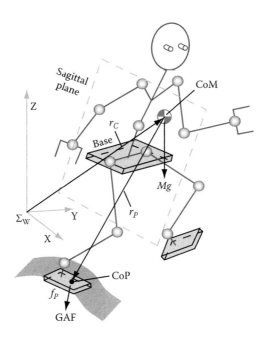

FIGURE 8.1 Definition of coordinate.

In Section 8.4, however, we focus on the multiple point constraint rather than orientation constraint because this representation is intuitive and flexible to negotiate various ground shapes. More specifically, we consider $\lambda = f_S = [f_{S1}, f_{S2}, \ldots, f_{S\alpha}]^T$, where $f_{Si} = [fx_{Si}, fx_{Si}, fx_{Si}]^T$ is the contact force at each contact point, that meets $\Sigma_1^\alpha f_{Si} = f_R$. Note that the orientation constraint can be expressed by a set of contact position constraints, for example, $z_{S1} = z_{S2} = 0$ and so on. The associated Jacobian $E(q)$, anyway, automatically relates its constraints into the contact forces. Overconstrained cases may happen, but are solved numerically in the optimization step.

Whichever constraint is used, this can be calculated as

$$f_p = (EI^{-1}E^T)^{-1} \left\{ \gamma + EI^{-1}(u - C - G) \right\} \tag{8.3}$$

which is obtained by solving Equations (8.1) and (8.2) for λ.

8.1.2 REPRESENTATION OF DYNAMICS IN TERMS OF COM AND GRF

Although full dynamics is derived in the above section, it is inconvenient to capture the gross behavior of the robot. As is well known from multidegree rigid body dynamics, the center of mass (COM) plays an important role. Here we introduce the COM position instead of that of the base, and rewrite the dynamics using these new generalized coordinates. By doing so we can obtain a useful dynamics representation suitable for controller derivation in the next section.

Let us begin with the simplest contact situation depicted in Figure 8.1. Let $r_C = [x_C, \ y_C, \ z_C]^T \in \mathcal{R}^3$ be the position vector of COM in the world coordinate frame Σ_W, and $r_P = [x_P, \ y_P, \ z_P]^T \in \mathcal{R}^3$ be the position vector from COM to the contact point. We assume the robot foot contacts the ground, and the center of pressure (COP) locates inside the supporting region (SR) composed of the vertex of the contact points. That is, the base orienatation ϕ and joint angle θ meet

$$\dot{\phi} = A(\phi, \theta)\dot{\theta} \qquad (8.4)$$

Then, we redefine the generalized coordinate as $q = [r_C, \theta]^T \in \mathcal{R}^{3+n}$.

Next, we assume the arbitrary contact point r_P in SR does not move:

$$r_C + r_P = 0 \qquad (8.5)$$

Combined with (8.4) we obtain

$$\dot{r}_C + J_P(\phi, \theta)\dot{\theta} = \underbrace{\left[id_3 \middle| J_P(\phi, \theta) \right]}_{E(\phi,\theta)} \dot{q} = 0 \qquad (8.6)$$

where $id_N \in \mathcal{R}^{N \times N}$ denotes the identity matrix of dimension N, and $J_P(\phi, \theta) = \partial r_P / \partial \theta \in \mathcal{R}^{3 \times n}$ is the contact Jacobian.

The readers may wonder where the constraint (reaction) moment associated with (8.4) appears. The constraint moment is derived in (8.3), but deleted in (8.10). This is because we determine the joint torque so that the COP lie in the SR. We assume the foot is rigid, as long as COP lies in SR, therefore this means the necessary reaction moment is automatically generated on every point on the foot. This point is made clear in a simple example described in Section 8.2. That is why we can consider the ground applied force $(GAF)f_p$ only which is applied to COP. If the foot is flexible enough, however, this is equivalent to considering the point contact foot. In this case, the reaction moment is supposed to be close to zero, and we have to think of an underactuated case, which is described in Section 8.2.2.

The reaction moments determined by specifying the COP are the pitch and roll moments only. The yaw moment also must be resolved in another way. We can explicitly specify the amount of the yaw moment and solve this by multiple constraint forces. *

In this setup, (8.1) is converted to the following decoupled dynamics,

$$\underbrace{\begin{bmatrix} M & 0 \\ \hline 0 & I(\phi, \theta) \end{bmatrix}}_{I} \underbrace{\begin{bmatrix} \ddot{r}_C \\ \hline \ddot{\theta} \end{bmatrix}}_{\ddot{q}} + \underbrace{\begin{bmatrix} 0 \\ \hline C(\phi, \theta, \dot{\theta})\dot{\theta} \end{bmatrix}}_{C} + \underbrace{\begin{bmatrix} -Mg \\ \hline 0 \end{bmatrix}}_{G} = \underbrace{\begin{bmatrix} 0 \\ \hline \tau \end{bmatrix}}_{u} - \underbrace{\begin{bmatrix} id_3 \\ \hline J_P(\phi, \theta)^T \end{bmatrix}}_{E^T} f_P \qquad (8.7)$$

wherein $M = \text{diag}(m, m, m) \in \mathcal{R}^{3 \times 3}$ is the mass matrix (m is the total mass); $I(\theta) \in \mathcal{R}^{n \times n}$ is the inertia matrix; and $C(\theta, \dot{\theta}) \in \mathcal{R}^n$ is the centrifugal and Coriolis term.

---

* In [29] we assumed q includes the orientation of the base. This is justified when the foot is fixed to the ground, and we assumed this situation for deriving the balancing controller.

The Jacobian $J_P(\phi, \theta)$ can be derived easily using a standard kinematics computation (See Reference [2] Section 3.1.3, e.g.). The upper part of (8.7) portrays simple linear dynamics:

$$M\ddot{r}_C = Mg + f_R \tag{8.8}$$

This shows the obvious fact that we can control $\ddot{r}_C$ by f_R, which is shown in Section 8.10.

Differentiating (8.6) yields

$$E(q)\ddot{q} + \gamma(q, \dot{q}) = 0 \tag{8.9}$$

where $\gamma(q, \dot{q}) = \partial/\partial q (E\dot{q})\dot{q}$. Solving (8.7) and (8.9) for $\ddot{q}$ results in

$$f_p = (EI^{-1}E^T)^{-1} \left\{ \gamma + EI^{-1}(u - C - G) \right\} \tag{8.10}$$

Therefore, the GAF is affected by the control input u, as well as the positions and velocities.

8.2   MODEL-BASED CONTROL: DYNAMIC BALANCING

This section introduces a model-based motion control for a full-body humanoid robot. Stabilization under unknown external disturbances can only be possible with feedback information. The controller naturally takes the feedback form of the task-space variables. The COM would be the most suitable task variable because it represents the gross behavior of the multibody dynamics compared with other controlled variables such as joint angle, posture, height of the body and so on. Therefore, the control objective for balancing is very simple: to stabilize the COM to some particular stable equilibrium points. The formulation described in the last section facilitates the derivation of the feedback controller in terms of the COM.

Linear or angular momentum is also a useful variable but could be optional for the balancing task because the former is a scaled velocity of the COM, and the latter could also be replaced by more intuitive variables. The latter is less intuitive, but its role in the balancing problem is discussed in the literature [26]. In the following, we treat the COM control first, then discuss the benefit of controlling angular momentum in Section 8.2.3.

8.2.1   DYNAMIC BALANCING 1: FULL-ACTUATED CASE

The balance control is asymptotic stabilization of the ground projection of the COM to some target position in the supporting convex hull. Controlling the COM height is optional. From the simple linear dynamics (8.8), it is easy to see that a linear feedback law with gravity compensation can be used:

$$\overline{f}_R = -Mg - K_{PC}(r_C - \overline{r}_C) - K_{DC}(\dot{r}_C - \dot{\overline{r}}_C) \tag{8.11}$$

with some desired COM position/velocity $\overline{r}_C$, $\dot{\overline{r}}_C$.

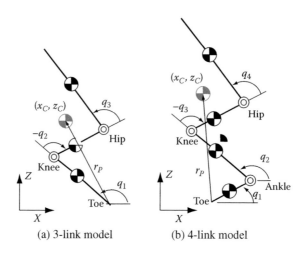

(a) 3-link model   (b) 4-link model

FIGURE 8.2   Planar under-actuated models.

Let us show a simple example with a planar humanoid model Fig. 8.2a. Equation (8.10) can be rewritten as

$$\underbrace{EI^{-1}(u - C - G)}_{A} = \underbrace{-\gamma + (EI^{-1}E^{T})f_p}_{B} \tag{8.12}$$

or

$$\left[\begin{array}{c|c} A_{11} & A_{12} \\ \hline A_{21} & A_{22} \end{array}\right] \left[\begin{array}{c} MG \\ \hline \tau - C\dot{q} \end{array}\right] = \left[\begin{array}{c} B_1 \\ \hline B_2 \end{array}\right] \tag{8.13}$$

Arranging the equation for τ, we obtain

$$\left[\begin{array}{c} A_{12} \\ A_{22} \end{array}\right] \tau = \left[\begin{array}{c} B_1 - A_{11}MG - A_{12}C\dot{q} \\ B_2 - A_{21}MG - A_{22}C\dot{q} \end{array}\right] \tag{8.14}$$

The size of the left coefficient matrix is 2×4, and we can solve the equation for τ by using a pseudo-inverse. Although it is not easy to solve the singularity, the above control torque exactly achieves $f_P = \bar{f}_P$ when the state $(q, \dot{q})$ is within the regular region.

Figure 8.4 shows the balancing control simulation of the three-link model (Figure 8.2a). The link parameters are set similar to the humanoid robot described in Section.

8.2.2 DYNAMIC BALANCING 2: UNDERACTUATED CASE

The above control scheme can be applied even in the single-point contact case, such as toe or heel contact. The single contact point is exactly the center of pressure because no other contact forces exist. In this case, the robot is underactuated because the robot cannot generate the constraint moment w_R around the contact point, and this must be

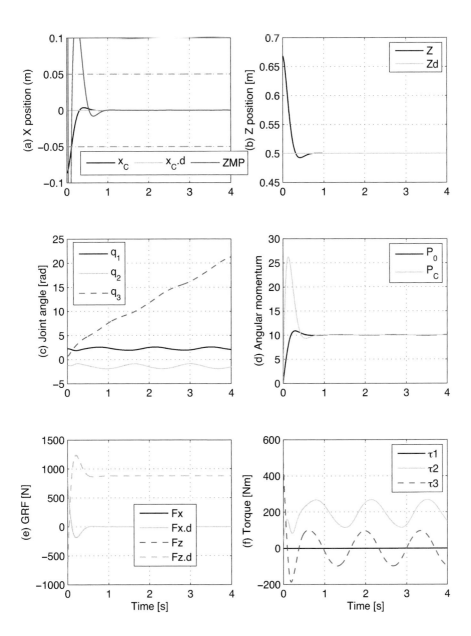

FIGURE 8.3   Three-link model underactuated balancing simulation. Note that the ankle torque is zero. That is, the COP is always located at ORG, whereas ZMP is not. As before, the actual GRF is exactly the same as the desired values.

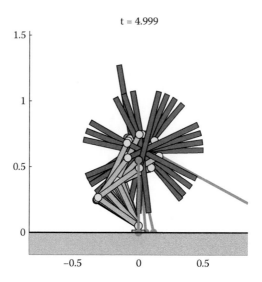

FIGURE 8.4

zero. The foot can rotate around the heel or toe. However, the robot can control the COM to some extent as long as the translational GAF is still available.

Suppose the robot is in toe contact. To give τ_1 freely (including zero), we separate it out from (8.14); then we obtain

$$
\begin{bmatrix} A_{12}(2:4) \\ A_{22}(2:4) \end{bmatrix}
\begin{bmatrix} \tau_2 \\ \tau_3 \\ \tau_4 \end{bmatrix}
=
\begin{bmatrix} B_1 - A_{11}MG - A_{12}C\dot{q} - A_{12}(1)\tau_1 \\ B_2 - A_{21}MG - A_{22}C\dot{q} - A_{22}(1)\tau_1 \end{bmatrix}
\tag{8.15}
$$

Note that this procedure is nothing more than the partial feedback linearization as shown in [3].

The size of the left coefficient matrix is 2×3, and we can solve the equation for (τ_2, τ_3, τ_4) by using a pseudo-inverse with some weighting matrix.

As long as the state evolves within the regular region, the control torque exactly achieves $f_P = \bar{f}_P$ for given τ_1.

Figure 8.5 and 8.6 shows the balancing control simulation of the same three-link model (Figure 8.2a) with $\tau_1 = 0$. Because the initial COM is far from the equilibrium, the robot has to move its joints very fast, while avoiding the singularity. We noticed that singularity occurs when the knee joint meets $q_2 \geq 0$. We use a simple joint limit (spring-damper) to avoid the singularity in this simulation at the cost of temporal error in the GRF. Systematic singularity avoidance control can be designed, but it is generally very difficult.

As one can see from the figure, the hip joint rotates unboundedly, and the COM stays at the target position. This shows the *zero-dynamics*, uncontrolled mode of the feedback-linearized system. Both angular momenta (AM) P_0 around the contact point or the one P_C around the COM are converging to constants (see Graph (d)).

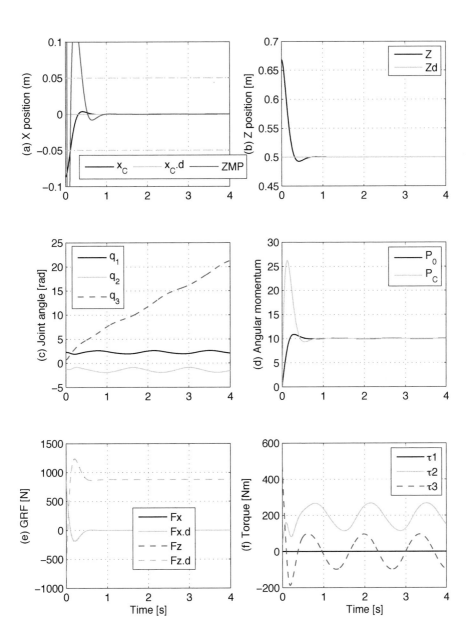

FIGURE 8.5 Three-link model underactuated balancing simulation.

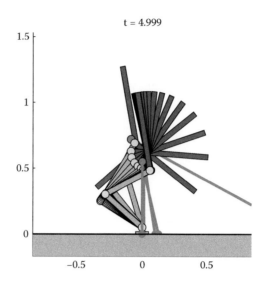

t = 4.999

FIGURE 8.6

Note that in Reference [3] a desired AM is specified to some value (zero), then its time-derivative together with (8.15) is solved. Instead of doing this, we show a much simpler solution below.

8.2.3 STABILIZATION OF ANGULAR MOMENTUM BY INTEGRAL ACTION

As we have seen earlier, merely stabilizing the COM does not solve the internal motions. This section briefly shows one of the remedies.

Let's take a look at the equation associated with AM in the planar robot model. That is, the third line of the equation of motion (8.7),

$$I_1 \ddot{q} + C_1 \dot{q} = \tau_1 + J_{P1}^T f_P \tag{8.16}$$

where I_1 and C_1 are the third row vector of the inertia matrix I and colinear matrix C, respectively (neither includes the cyclic variable q_1), and $J_{P1} = [-z_C, \ x_C]^T$ is the submatrix of $J_P = [J_{P1}|J_{P2}]$. Because the AM around the COM is represented by $P_C = I_1 \dot{q}$, (8.16) can be rewritten as

$$\dot{P}_C = \tau_1 + J_{P1}^T f_P \tag{8.17}$$

On the other hand, the AM around the contact point is expressed, by definition, as

$$P_0 = P_C + r_C \times m\dot{r}_C \tag{8.18}$$

and its time-derivative is given by

$$\begin{aligned} \dot{P}_0 &= \dot{P}_C + m(z_C \ddot{x}_C - x_C \ddot{z}_C) \\ &= \dot{P}_C + z_C f_x - x_C(f_z - mg) \end{aligned} \tag{8.19}$$

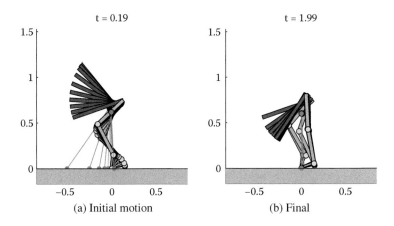

$t = 0.19$

$t = 1.99$

(a) Initial motion

(b) Final

FIGURE 8.7 Animation corresponding to Figure 8.8.

Substituting (8.17) into (8.19) yields

$$\dot{P}_0 = (-z_C f_x + x_C f_z + \tau_1) + z_C f_x - x_C(f_z - mg)$$
$$= \tau_1 + x_C mg \tag{8.20}$$

From (8.18) we reach the following statement. Even if the COM is asymptotically stable $r_C \times m\dot{r}_C \to 0$, it can happen that $P_0 \neq 0$ if $P_C \neq 0$. This is the internal motion which we have in Section 8.2.2.

Consequently, from (8.20) it is found that we should use

$$f_{px} = -K_P x_C - K_D \dot{x}_C - K_I \int (\tau_1 + x_C mg)dt \tag{8.21}$$

instead of (8.11), to regulate AM. Thus we can achieve both $\dot{r}_C \to 0$ and $P_0 \to 0$, which turns out that both linear and angular momenta become zero; that is, all motion stops.

Figures 8.7 and 8.8 shows the result of the proposed controller (8.21). The conditions are exactly the same as in the previous simulation (Figure 8.5) except for the additional integral action. One can see the internal motions completely disappear.

Similar balancing simulations are conducted using the four-link model with toe contact (Fig. 8.2b). The result is shown in Figures 8.7 and 8.8, where the weight matrix is adopted. Interestingly, the recovery motion looks like humans behavior. The COM and ZMP rapidly approach the origin (toe) within 0.2 s as shown in Figure 8.7a, and then slowly converge* as shown in Figure 8.7b. Note that no torque is applied around the toe (see $\tau_1 = 0$ in Graph (f)). The AM rapidly increases from the initial value (zero) as the COM is pulled back to the origin, and then converges to zero as the internal motion (joint motion) converges. The global error response is determined by a linear feedback controller (8.21). Note there is a discontinuous change around

---

* It would be natural to set the COM target to a position a bit behind the toe. Then, the robot can relax at the static equilibrium.

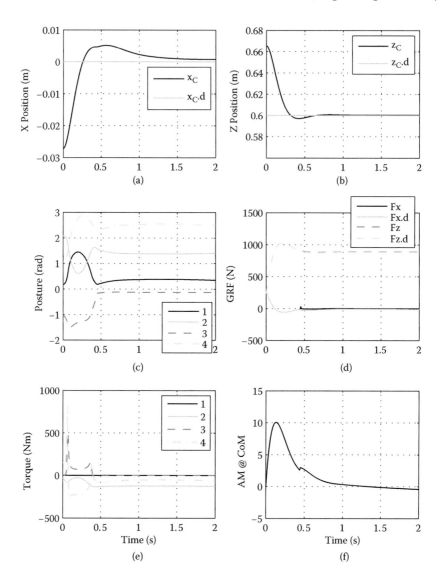

FIGURE 8.8   Four-link model underactuated balancing simulation.

0.45 s. This is because of the knee joint hitting its limit. Because we use a simple spring-damping model for the limit, there are small errors in the GRF (Graph (d)). The torso is largely tilted from the initial upright posture. It is expected that the torso tilt could be reduced if we put the arms to the torso.

The actual region of attraction is limited by physical constraints such as maximum torque, range of joint angle, and the like. It is, however, difficult to obtain directly the optimal joint torque required for balancing in general. Although some analysis on a simple pendulum model with a flywheel is useful [4], it is generally difficult to compute online the optimal joint torque required for dynamic balancing for all joints.

8.3 PRACTICAL SOLUTION TO REAL-WORLD PROBLEMS: PASSIVITY-BASED CONTACT FORCE CONTROL FRAMEWORK

The controller derived above works in simulation, but it is hard to believe humans adopt such computational processes for balancing tasks. The reason is that it requires an exact model including the inertia matrix, which becomes impractical for large DoF systems. Second-ordered velocity terms are especially problematic because the velocity measurement is usually noisy. The controller structure has extremely complicated parameterization that involves many inverse operations. Now we have to think about how to solve the problem practically. This process naturally let us challenge robust humanoid motor control.

This section introduces our passivity-based approach which first appeared in [29]. Under some natural assumptions, we show that a simple controller can handle the ground reaction force, hence balancing. We also introduce a multicontact force control problem, which is crucial for compliant motion control for humanoid robots. We experimentally validate the methods on a real full-size humanoid robot introduced in Section X.

8.3.1 GRAVITY COMPENSATION WITH DAMPING INJECTION FOR CONTACT FORCE CONTROL

We continue the discussion in Section 8.2. Now, the problem is how to transform the desired contact force into joint torques "effectively." Since dynamic nonlinear term is difficult to identify and utilize, therefore we start with the static case. This simple idea leads us to gravity compensation (G-comp) for contact force control.

Suppose the robot foot fully contacts the ground. Therefore, the first three entries in τ are available; that is, the robot is fully actuated as we considered the case in Section 8.2. In this case, $\dot{q} \approx 0$, therefore $\gamma \approx 0$, $C \approx 0$ hold. For some new force input $f_u = [f_{ux}, \ f_{uy}, \ f_{uz}]^T \in \mathcal{R}^3$, it is straightforward to see that the joint torque

$$\tau = J_P^T(f_v + Mg) \tag{8.22}$$

with

$$f_v = f_u + (J_P I^{-1} J_P^T)^{-1} M^{-1} f_u \tag{8.23}$$

renders the closed-loop system satisfying

$$f_P \approx f_u + Mg \tag{8.24}$$

This is the main idea of G-comp for controlling the contact force. Then, $M\ddot{r}_C \approx -f_u$, meaning that we can control the COM movements using f_u. The gravity term is readily identifiable and a dominant nonlinear term in full dynamics. As shown below, G-comp plays a central role not in balancing but also in natural interaction with the environment.

8.3.2 PROOF OF EQUATION (8.24)

By substituting (8.22)

$$\tau = J_P^T(f_v + Mg)$$

into

$$f_P \approx (EI^{-1}E^T)^{-1}\{EI^{-1}(u - G)\} \tag{8.25}$$

and expanding the right-hand side, we obtain

$$EI^{-1}E^T$$
$$= [id | J_P] \begin{bmatrix} M & 0 \\ 0 & I \end{bmatrix}^{-1} \begin{bmatrix} id \\ Jp^T \end{bmatrix} \tag{8.26}$$
$$= M^{-1} + J_P I^{-1} J_P^T$$

and

$$EI^{-1}(u - G)$$
$$= EI^{-1}\left(\begin{bmatrix} 0 \\ \hline Jp^T(Mg + f_v) \end{bmatrix} + \begin{bmatrix} Mg \\ 0 \end{bmatrix} \right)$$
$$= EI^{-1}E^T Mg + EI^{-1} \begin{bmatrix} 0 \\ Jp^T f_v \end{bmatrix} \tag{8.27}$$
$$= EI^{-1}E^T Mg + \left(J_P I^{-1} J_P^T \right) f_v.$$

With reference to (8.26) and (8.27), (8.25) becomes

$$f_P \approx Mg + \left(M^{-1} + J_P I^{-1} J_P^T \right)^{-1} \left(J_P I^{-1} J_P^T \right) f_v \tag{8.28}$$

If we give f_v as (8.23), that is,

$$f_v = \left(J_P I^{-1} J_P^T \right)^{-1} \left(M^{-1} + J_P I^{-1} J_P^T \right) f_u$$
$$= f_u + \left(J_P I^{-1} J_P^T \right)^{-1} M^{-1} f_u$$

we recover (8.24):

$$f_P \approx f_u + Mg$$

In (8.23), we used the inertial matrix $I(q)$, which might be difficult to obtain in some applications. A simpler formula is available by setting

$$\tau = J_P^T(f_u + Mg) \tag{8.29}$$

that is, $f_v := f_u$, instead of (8.22) and (8.23). This approximation is valid whenever $f_u = 0$ (G-comp only) or f_u is given by some feedback law. We assume this is true for all ranges of motion. If the posture is singular this coefficient becomes very large and (8.24) does not hold unless $f_u = 0$.

On the other hand, the static force control for redundant manipulators is well known to cause internal motions [5]. Some dynamic effects might arise as the internal motions if γ and C are not small. We must therefore compensate for the internal motion somehow, while simultaneously achieving a desired GAF. One simple way to accomplish these requirements is to assign a simple nonlinear term that compensates γ and C. Using (8.22) and $\partial/\partial q (J_P \dot{q})\dot{q} = -J_P \ddot{q}$, we can derive a modified version of (8.22):

$$\tau = J_P^T (f_v + Mg) + \zeta(q, \dot{q}) \tag{8.30}$$

which yields convergence of the GAF:

$$f_P \rightarrow f_u + Mg \ (t \rightarrow \infty) \tag{8.31}$$

provided $\zeta(q, \dot{q})$ is designed so that

$$I(q)\ddot{q} + C(q, \dot{q}) - \zeta(q, \dot{q}) = 0 \tag{8.32}$$

is stable.

8.3.3 PROOF OF EQUATION (8.31)

Substituting $\tau = J_P^T Mg + \zeta$ into (8.10) yields

$$\gamma_2 + J_P I^{-1}(\zeta - C) =: w \tag{8.33}$$

where γ_2 is the lower half of $\gamma = [\gamma_1, \ \gamma_2]^T$. On the other hand, the lower part of (8.9)

$$E(q)\ddot{q}_C + \gamma(q, \dot{q}) = 0$$

can be written as

$$J_P \ddot{q} + \gamma_2 = 0 \tag{8.34}$$

Combining (8.33) and (8.34) yields

$$w = -J_P I^{-1} \left\{ I\ddot{q} + C - \zeta \right\} \tag{8.35}$$

Therefore, if we design ζ so that $I\ddot{q} + C - \zeta = 0$ is met (exact nonlinear compensation) or $I\ddot{q} + C - \zeta$ is asymptotically stable, that is, $\dot{q} \rightarrow 0$ as $t \rightarrow \infty$ (e.g., via damping injection), then $w \rightarrow 0$ as $t \rightarrow \infty$. Combining this with the result from the above proof we obtain (8.31)

$$f_P \rightarrow f_u + Mg \quad (t \rightarrow \infty)$$

As a conservative solution we can simply set ζ as jointwise damping:

$$\zeta = -D\dot{q} \tag{8.36}$$

with a constant diagonal matrix $D > 0$. Although we should investigate the structure of the left-hand side of (8.32), the controller (8.30) itself can be made very simple, as in this example. See References [6] [7] for a damping injection strategy for controlling redundant manipulators, where the authors analyzed the convergence of the internal motions and the position error of the end-effector.

The closed-loop system, becomes

$$M\ddot{r}_C = -f_u - f_\zeta + f_E \tag{8.37}$$
$$I(q)\ddot{q} + C(q,\dot{q}) = J_P(q)^T f_v + \zeta(q,\dot{q}) \tag{8.38}$$

where f_ζ is the dissipation term associated with the third term of (8.30). As a result, we obtain the *passivity* [8] [9] relationship

$$H(q,\dot{q}) := \frac{1}{2}\dot{r}_C^T M\dot{r}_C + \frac{1}{2}\dot{q}^T I\dot{q} = \frac{1}{2}\dot{q}^T \left(J_P^T M J_P + I\right)\dot{q}$$
$$\leq \int f_v^T J_P(q)\dot{q}dt + \int f_E^T J_P(q)\dot{q}dt \tag{8.39}$$

Consequently, we have the following properties:

- COM tracking is possible with the addition of simple controllers using f_u as in Section 8.10.
- Without external forces f_E and the damping term, the COM is subject to the constant velocity movement.
- Without new force input f_u, the COM and all the joint movements can be driven externally by f_E,

8.3.4 PASSIVITY-BASED DYNAMIC BALANCING

Recall that the COP is, by definition, the representative force application point because it is the weighted sum of the translational ground contact forces as in (8.45). On the other hand, the *zero moment point* (ZMP) [10] is defined as a point on the ground at which total moments balance and references therein, which is denoted as $r_P' = [x_P',\ y_P',\ z_P]$. By the original definition, ZMP can be computed as

$$x_P' = \frac{z_P m\ddot{x}_C + \dot{P}_{cy}}{m(\ddot{z}_C + g)}$$
$$y_P' = \frac{z_P m\ddot{y}_C + \dot{P}_{cx}}{m(\ddot{y}_C + g)} \tag{8.40}$$

Note that there is a time derivative of angular momentum around the COM in these equations. See the discussion in Section 8.2.

Inasmuch as we do not use complex nonlinear terms in the passivity formulation, a "simple version" of ZMP is introduced. Specifically, we assume the rate change of the angular momentum around the COM is negligible. In this case, ZMP can be approximated as

$$
\begin{aligned}
x'_P &= \frac{z_P m \ddot{x}_C}{m(\ddot{z}_C + g)} = \frac{z_P f_{xP}}{f_{zP}} \\
y'_P &= \frac{z_P m \ddot{y}_C}{m(\ddot{y}_C + g)} = \frac{z_P f_{yP}}{f_{zP}}
\end{aligned}
\tag{8.41}
$$

Note that this approximated ZMP coincides with the COP "as long as" it lies inside SR (because the COP cannot exit from SR).

Therefore, with our G-comp strategy (8.29) and new force input (8.11), the desired ZMP $\bar{r}'_P$ is set to:

$$
\bar{x}'_P = \frac{z_P f_{ux}}{mg + f_{uz}}, \qquad \bar{y}'_P = \frac{z_P f_{uy}}{mg + f_{uz}}
\tag{8.42}
$$

In those equations, $z_P = -z_C$. We restrict ZMP into the supporting convex hull to incorporate the balancing controller with our contact force distribution about the COP. We can modify the horizontal GAF (f_{ux} and f_{uy}) and/or normal GAF (f_{uz}) to do that. We use this as the desired COP for balancing. That is, we set

$$
\bar{r}_P = \bar{r}'_P
\tag{8.43}
$$

Finally, the modified GAF $\bar{f}_P$, in addition to the desired COP $\bar{r}_P$, are then used for determining the desired contact force closure in Section 8.4. Note that the approximated ZMP can still exit from the supporting region depending on the actual GRF.

An intuitive interpretation is possible by considering a special case: $\bar{f}_{xP} = \bar{f}_{yP} = 0$ (G-comp only). In this case, the desired ZMP (or COP) coincides with the ground projection of the COM, $\bar{x}_P = \bar{y}_P = 0$, if they are inside the supporting region. Because the gravity is compensated the COM exhibits approximately constant velocity movement (when the damping is small enough).

Figure 8.9 shows the simulation result. The initial condition of gCOM cannot be set outside SR.

8.4   HANDLING MULTIPLE CONTACT POINTS

The GAF controller derived in (8.30) is insufficient for humanoid robots interacting with the environment at multiple contact points. If the robot is in the double support phase or quadruped phase, there are many possible combinations of the commanded joint torques that result in the same desired GAF. However, it is not easy to solve for the required joint torques because of multiple kinematic loops [11].

We adopt a two-stage solution: that is, determine the force closure first explicitly, then transform it to the joint torques. The latter was already done with the passivity-based framework in the the previous section. The main problem in this section, therefore, is how to distribute the desired GAF to multiple contact points in optimal ways.

In this section, we show a simple solution suitable for compliant humanoid robots; that is the main objective of this chapter. The method has two features:

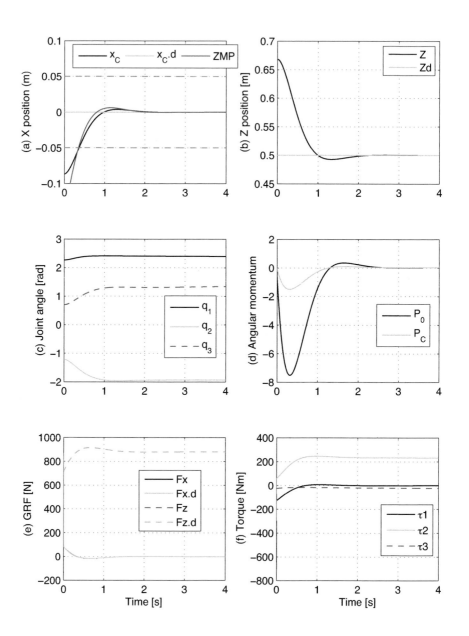

FIGURE 8.9   Passivity-based full-actuated balancing simulation on three-link robot model. The robot can balance only if the COM is inside SR, and the balancing ability is limited.

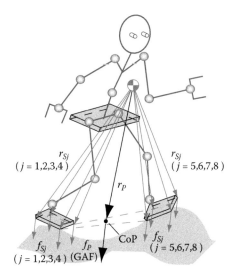

FIGURE 8.10   Definition of contact points and forces: each ground contact point r_{Sj} is assigned the contact force f_{Sj}, and each nonground contact point r_{Fj} is assigned the interaction force f_{Fj}. The contact forces f_{Sj} are determined using a desired GAF f_P and f_{Fj}. The COP r_P always lies within the supporting convex hull of r_{Sj}.

- To treat force control in a task space in an intuitive and uniform manner, we first solely consider the translational forces of the active force application points.
- To make the algorithm simple, we first consider colinear contact points to use linear programming.

The latter does not mean that we ignore the general contact situation. The general solution has been considered in the grasping literature (e.g., quadratic programming for convex optimization). However, in the biped balancing context, it is reasonable to assume the robot is walking on flat ground. We use force control, therefore the foot is compliantly adapted to the ground shape. Gravity and gravity compensation play important roles. For the largely deformed shape, we can again intuitively organize the contact forces appropriately. The detailed discussion is given in Section 8.7.

8.4.1 Optimal Contact Force Distribution

Assuming that we are interested in a total of α contact points defined as $r_S = [r_{S1}, r_{S2}, \ldots, r_{S\alpha}]^T \in \mathcal{R}^{3\alpha}$, where each vector component is $r_{Si} = [x_{Si}, y_{Si}, z_{Si}]^T \in \mathcal{R}^3$ for each index i, the associated applied contact forces are defined as $f_S = [f_{S1}, f_{S2}, \ldots, f_{S\alpha}]^T \in \mathcal{R}^{3\alpha}$, where each vector component is $f_{Si} = [x_{Si}, y_{Si}, z_{Si}]^T \in \mathcal{R}^3$. In this description, f_S is the force that the robot applies to the contact points (care is needed in the force direction). Of course, COP r_P must lie within the supporting convex hull of r_{Sj}. We are also interested in the total of β nonground contact points or interaction points as $r_F = [r_{F1}, r_{F2}, \ldots, r_{F\beta}]^T \in \mathcal{R}^{3\beta}$ and their associated applied forces $f_F = [f_{F1}, f_{F2}, \ldots, f_{F\beta}]^T \in \mathcal{R}^{3\beta}$. For example, $\alpha = 8$ and $\beta = 8$ in Figure 8.13a, but $\alpha = 4$ and $\beta = 12$ in Section 8.4.

The discussion here is independent of robot models, and the number and location of the contact points are arbitrary; they are applicable to one-legged, bipedal, and quadrupedal robots, as long as the COP is definable. It should be noted that the contact points do not necessarily lie in the same plane.

The main task is to distribute the GAF to each ground contact point through the COP. From the definition, the COP can be expressed as

$$x_P = \frac{\sum_{j=1}^{\alpha} x_{Sj} f_{zSj}}{\sum_{j=1}^{\alpha} f_{zSj}}, \quad y_P = \frac{\sum_{j=1}^{\alpha} y_{Sj} f_{zSj}}{\sum_{j=1}^{\alpha} f_{zSj}} \tag{8.44}$$

and the GAF is the sum of the respective contact forces:

$$f_P = \sum_{j=1}^{\alpha} f_{Sj} \tag{8.45}$$

In simple notation, these two equations combine to

$$\begin{bmatrix} x_P \\ y_P \\ 1 \end{bmatrix} f_{zP} = \underbrace{\begin{bmatrix} x_{S1} & x_{S2} & \cdots & x_{S\alpha} \\ y_{S1} & y_{S2} & \cdots & y_{S\alpha} \\ 1 & 1 & \cdots & 1 \end{bmatrix}}_{A_z \in \mathcal{R}^{3\times\alpha}} \begin{bmatrix} f_{zS1} \\ f_{zS2} \\ \cdots \\ f_{zS\alpha} \end{bmatrix} \tag{8.46}$$

where A_z represents a contact-force distribution matrix.

Given a desired normal GAF $\overline{f}_{zP}$ and COP $\overline{r}_P$ (see Section 8.10 for the balancing control case), we can calculate the corresponding desired normal contact forces $\overline{f}_{zS}$. Specifically, we propose an optimal contact force distribution calculated as

$$\begin{bmatrix} \overline{f}_{zS1} \\ \overline{f}_{zS2} \\ \cdots \\ \overline{f}_{zS\alpha} \end{bmatrix} = A_z^{\#} \begin{bmatrix} \overline{x}_P \\ \overline{y}_P \\ 1 \end{bmatrix} \overline{f}_{zP} \tag{8.47}$$

with $A_z^{\#} = (A_z^T A_z)^{-1} A_z^T$. This solution is optimal in the sense that it minimizes the sum of the total contact forces. Because of the convexity of the contact points, the resultant normal contact force $\overline{f}_{zS}$ cannot be positive (therefore, the unilateral constraint condition is met) as long as $\overline{f}_{zP}$ is negative and the desired COP $\overline{r}_P$ is set within the supporting convex hull (they are obvious restrictions).* Similarly, the horizontal contact forces are optimally distributed.

---

* As a trivial extension, we can introduce a weight into the optimal criteria as $\overline{f}_{zS}^T W \overline{f}_{zS}$, where $W \in \mathcal{R}^{\alpha\times\alpha}$ is the weighting matrix. The optimal force closure is given as (8.47), but with $A_z^{\#} = (A_z^T W^{-1} A_z)^{-1} A_z^T W^{-1}$. The weighting might be useful when we need to consider the friction cone of the ground explicitly, or when we need to reduce the load to some limbs, for example, in case of breakage.

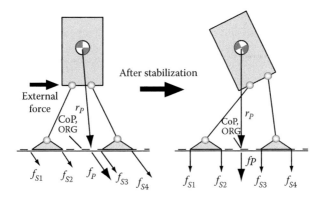

FIGURE 8.11   Adaptation to external forces. Balance recovery force keeps the COM to its origin (ORG), but allows joints to deviate from their original posture.

Finally, the commanded joint torques are obtainable as

$$\tau = J_S(q)^T \overline{f}_S + \zeta(q, \dot{q}) \tag{8.48}$$

where $J_S(q) \in \mathcal{R}^{3\alpha \times n}$ represents the contact Jacobian from the COM to the supporting contact points (derivative of r_S with respect to q), where G-comp $\overline{f}_P = f_v + Mg$ must be included in the calculation of $\overline{f}_S$.

Figure 8.11 provides a schematic explanation of adaptation to unknown force disturbances. Suppose the external force is applied to the waist when the robot is in an upright equilibrium configuration just before the push. Based on the COM state, the feedback controller (8.11) applies GAF to counteract the disturbance, as shown in the left figure. As a result, the robot can achieve a new equilibrium configuration, as in Figure 8.11 on the right. With this simple solution, compliant (torque-controlled) balancing on a full-size humanoid robot is first achieved, which is shown in Section 8.5.

8.4.2   REACHING AND FORCE-INTERACTION TASKS

Using the above simple contact force framework, the motion command for supporting leg joints or swinging leg joints can be handled in a uniform manner. Contact forces at the interested contact points are used for task-space force control, and also for position tracking. For example, we can use the interaction forces f_{Fj} for controlling the position r_{Fj} of the swinging legs or hands using a simple PD feedback law

$$\overline{f}_{Fj} = -K_{PF}\left(r_{Fj}^W - \overline{r}_{Fj}^W\right) - K_{DF}\left(\dot{r}_{Fj}^W - \dot{\overline{r}}_{Fj}^W\right) \qquad (j = 1, 2, \ldots, \beta) \tag{8.49}$$

for given desired positions and velocities $\overline{r}_{Fj}^W = r_C + \overline{r}_{Fj}, \dot{\overline{r}}_{Fj}^W = \dot{r}_C + \dot{\overline{r}}_{Fj}$ in the world coordinate frame and positive gain matrices $K_P, K_D > 0$, or by a feedforward force command. The control torque is given simply as

$$\tau_f = J_F(q)^T \overline{f}_F \tag{8.50}$$

where $J_F(q) = \partial r_F / \partial q$ is the task Jacobian.

Even if the task control force is zero $\overline{f}_F = 0$, the swinging legs or hands are gravity-compensated and the internal motions are damped by the main full-body force controller (8.48). However, we must deactivate the contribution from balancing force input f_u to those joints $J_F(q)$ is involved in if we require that the reaching and force interaction task have the same priority as the balancing control. In this case, we can hold the swinging legs or hands and move them with a slight external force applied to arbitrary contact points.

To achieve good tracking of a desired GAF $\overline{f}_P$ the interaction forces must be compensated.* A simple way to achieve this is just subtracting the sum of the interaction forces from the desired GAF in advance:

$$\overline{f}_P \xleftarrow{\text{substitute}} \overline{f}_P - \sum_{j=1}^{\beta} \overline{f}_{Fj} \qquad (8.51)$$

However, this is an approximated compensation. Perfect compensation requires precise forward dynamics (moments of inertia identification and velocity measurement), referring back to the equation of motion, which is conceptually trivial, but computationally inconsistent with the above simple and practical framework.

8.5   EXPERIMENTS AND SIMULATIONS

This section provides selected results of the experiments and simulations. We use the life-size humanoid robot.

8.5.1   SIMPLE BALANCING EXPERIMENT

First we show the balancing experiments. One result is shown in Figure 8.16. Large disturbances were applied to the robot. Note that the robot is supported neither by hydraulic hoses nor by cables. They apply nonnegligible disturbances to the robot. There are two safety slings between the robot and the ceiling, which are slack except during emergencies.

The top two graphs show positions of the COM and desired COP. The limits on the desired COP are indicated by two dashed lines, ± 0.135 m. The desired COP always leads the COM, which reflects that the controller applies a GAF so that the COM converges to the desired position (zero in this example). The graphs in the third row depict the desired normal contact forces $\overline{f}_{zSj}$, as calculated using (8.47). The left graph shows the forces applied to the *front* corners (right sole, $f_{zS1} + f_{zS2}$; left sole, $f_{zS5} + f_{zS6}$; and the sum); the right graph shows the contact forces applied to the rear corners (right sole, $f_{zS3} + f_{zS4}$; left sole, $f_{zS7} + f_{zS8}$; and sum). The dashed line indicates the total antigravitational force: $-mg$. The COM height is not controlled in this example. Therefore, $\sum_{j=1}^{8} f_{zSj} = -mg$ holds. The graphs show that the antigravitational force is actually distributed over the normal contact forces appropriately by (8.47).

---

* If COM state feedback such as (8.11) is employed and the motions of legs/hands are sufficiently slow, the force/moment compensation is not necessary.

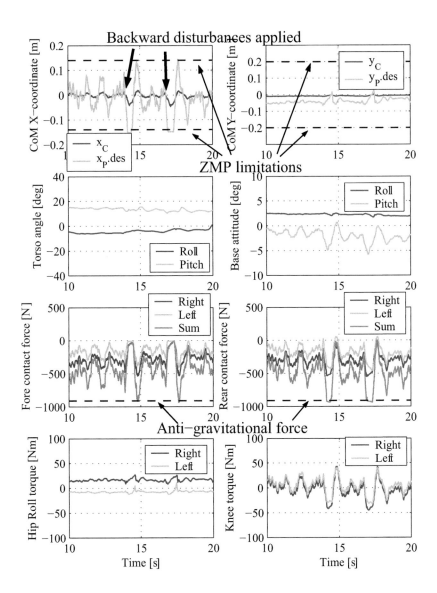

FIGURE 8.12   Experimental data of balancing disturbances.

We can test more aggressive motion control using simulation software. As an example, a periodic squat is shown in Figure 8.14 and Figure 8.15. The task feedback gain for (8.11) was set to $K_{PC} = 50$ N/m and $K_{DC} = 3$ Ns/m. The joint damping in (8.48) was set to zero for all limbs,* except for the torso joints: 5 Ns/m. The torso orientation is locally stabilized with a simple PD control. The desired COP is allowed

---

* The joints already have (unknown) physical damping due to hydraulic actuation.

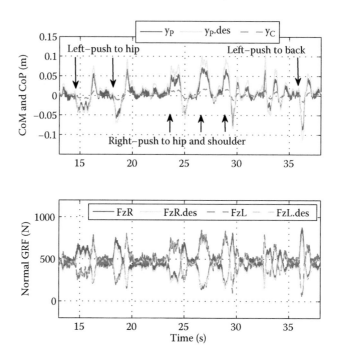

FIGURE 8.13   Experimental result of balancing. (Top) the COM is kept near zero, and the actual COP, estimated by normal ground reaction forces only, is roughly tracking the desired value. (Bottom) True GAF at each foot (FzR and FzL) is tracking the desired GRF (FzR.des and FzL.des).

to travel from -0.05 m to $+0.1$ m in the X-direction when both feet are aligned. The desired horizontal position of the COM is the center of the supporting convex hull, which is not fixed in Σ_W. The desired COM height $\bar{z}_C$ and the torso orientation are given simply by sinusoidal trajectories, as shown in Figure 8.15. In simulations, the gains in (8.11) are set to $(K_{PC}, K_{DC}) = (5,000, 500)$ for vertical motion and $(K_{PC}, K_{DC}) = (100, 50)$ for horizontal motion. The jointwise damping in (8.36) was set to $D = \text{diag}[d_1, d_2, \ldots, d_n]$ with $d_{legs} = 2$ and $d_{torso} = 8$.

FIGURE 8.14   Animation of a balanced squat with the torso swinging for 1–3 s: two red markers designate the COM and the ground projection of COM, whereas yellow and green markers demarcate the desired ZMP and actual ZMP.

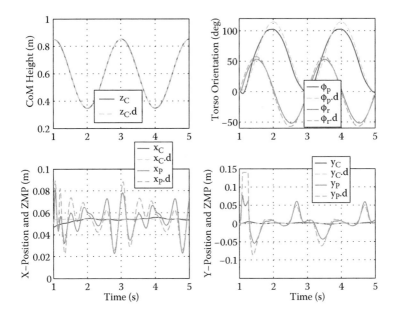

FIGURE 8.15 Time evolution of the balanced squat in Figure 8.14: COM (x_C, y_C, z_C) and ZMP (x_P, y_P) are shown along with their desired values (indicated by ".d") .

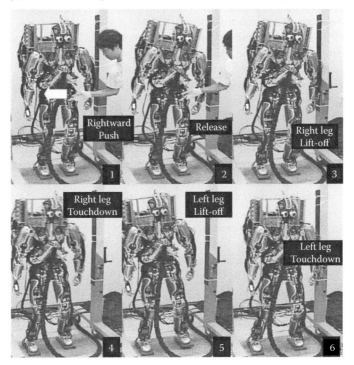

FIGURE 8.16 Our force distribution law reduces some contact forces when a large error in the COM is detected. As a result, a foot might lift off the ground.

We note that the joint torque control framework best suits the conventional multi-body dynamics simulator (software) because we can ignore the actuator dynamics and directly command joint torque inputs provided by the software. Once the joint torque servo control is established on the hardware side, the closed-loop behavior of the robot and the simulator is expected to be similar except for the environment dynamics such as contact dynamics. A contact model with torque-controlled robots is much more stable than that of position-controlled robots.

8.5.2 NATURAL FORCE INTERACTION AND REACHING TASK

The location to apply external force is arbitrary. Figure 8.16 shows the typical performance when a side force is applied. This result nicely illustrates the relationship between the COM error and the ground-reaction force distribution. In this example, the COM is largely laterally perturbed from the desired equilibrium (zero). The desired COP is moved onto one edge of the foot and the desired COM is fixed to the center. When this happens the desired normal contact forces of the other foot become nearly zero. For that reason, one foot loses its contact with the ground. In this example, however, we applied no position control for the swinging leg. Therefore, the lifted foot moves purely as a result of the balancing control. (See Section 8.8.1 for an active stepping control for push recovery.) In this case, the foot moves to the outside. Simultaneously, the COM moves back to the center of the feet. Then the lifted foot touches the ground. This process repeats once more; finally, the robot balances, but with a different configuration from the initial one.

Figure 8.17 shows the time evolution of the experiment. The top two graphs show the positions of the COM and the desired COP. The lifted foot does not move if we apply the feedback controller to maintain the distance between each foot.

A salient point is that we do not specify which joints should move in compliance with the disturbances. For example, we use no weighting matrix, as in some inverse kinematics approaches. The weighting is done automatically by the Jacobian transpose $J_S(q)^T$ in (8.48), and the compensating contact forces are distributed to the joint torques of the whole body. Therefore, if we push the robot torso, the torso compliantly follows first; then the lower body begins to compensate for the COM error. Similarly, if we push the hip, the hip moves first, as shown in Figure 8.12. We can even apply external forces to the leg joints. For example, if we push the knee, the knee joint follows compliantly to the external force; then the remaining joints generate compensatory torques to retain balance. Although the controller does not measure the external force, the response to the external force is immediate. The reason is that most joints are commanded to have zero torque when the robot has upright posture (equilibrium). The compliance is not only important for natural and safe human–humanoid interaction. It is also crucial to make the robot perform humanlike behavior.

A reaching task with balance is shown in Figure 8.18. The COM height z_C is not controlled, but is largely damped by the derivative gain K_{DC} in (8.11). Other parameters are identical to those given previously. The position of the swinging leg in Σ_W, r_F^W, is given as a sinusoidal function, as shown in Figure 8.19. During the motion, the target equilibrium point, that is, the center of the supporting foot, is moved slightly

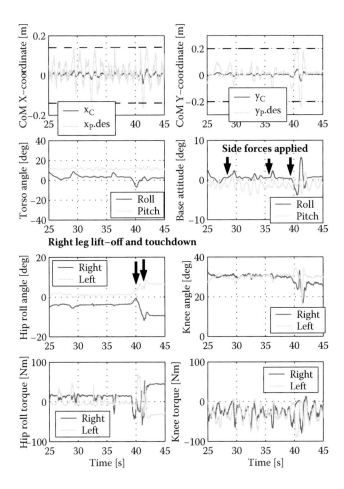

FIGURE 8.17   Experimental data corresponding to Figure 8.16.

because of the slip. Nevertheless, the horizontal COM position x_C, y_C is stabilized to the new equilibrium (the center of the sole). In both simulations, small tracking errors in the COM are apparent. The errors originate in the uncompensated dynamics and persist as long as the robot is in motion.

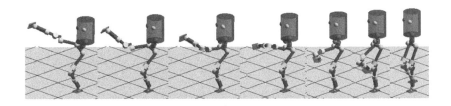

FIGURE 8.18   Animation of a periodic reaching task by a leg with simultaneous balancing.

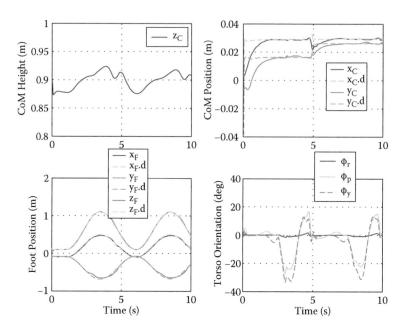

FIGURE 8.19 Simulation result of a reaching task by a leg. The target foot positions indicated by $x_F.d$, $y_F.d$, $z_F.d$ are given by 0.2 Hz sinusoidal functions. The horizontal position of COM x_C, y_C is regulated to the center of the foot.

8.6   COMPARISON WITH OTHER CONTROLLERS

Before proceeding further, it would be worth while to review the related works.

8.6.1   COMPARISON WITH ALTERNATIVE FORCE CONTROL SCHEME

The torque controllability of our robot relies on high-speed feedback from force sensors, not on (actuator model-based) feedforward control, as usually assumed in simulations. Here we consider our force control scheme from the viewpoint of force sensing and estimation.

The simplest force controller without a damping term is given by

$$\bar{\tau} = J_P^T \bar{f}_P \tag{8.52}$$

This controller does not measure the actual f_P (e.g., the known gravity Mg). In a static *case*, however, the actual joint torque is related to actual force by

$$\tau \approx J_P^T f_P \tag{8.53}$$

Suppose we use a simple proportional torque feedback controller

$$i = -K_t(\tau - \bar{\tau}) \tag{8.54}$$

where K_t is the torque feedback gain that might depend on each joint kinematics. Using (8.52) and (8.53) we obtain

$$i = -K_t J_P^T (\hat{f}_P - \overline{f}_P) \tag{8.55}$$

Therefore, in a sense, our torque feedback controller (8.54) can be considered an indirect force feedback controller with estimated force $\hat{f}_P$.

Although it is not our case, but if the control purpose is only controlling the task-space force and the force measurement is available, we can use (8.55); now $\hat{f}_P$ is replaced by the actual measurement and results in

$$i = -K_t J_P^T (f_P - \overline{f}_P) \tag{8.56}$$

This direct force feedback scheme, however, only works if K_t is high enough.

Otherwise, (8.56) does not provide full-body compliance because zero input ($i = 0$) stiffens the joints in a stiff actuator such as hydraulic cylinders controlled by flow-controlled servo valves or electric motors with high-reduction gears (not back-drivable). However, large K_t leads to a small force tracking error to the large actuator inputs, which may in turn cause larger force error due to the long transmission path from the actuator to the end-effector (feet).

8.6.2 COMPARISON WITH ALTERNATIVE POSITION CONTROL SCHEME

One can employ position-based force feedback controllers with $J_P^\#$ instead of J_P^T. Readers are recommended to refer to [2] for comparison between Jacobian transpose control and inverse control for task-space "position" control (not for force control). Historically, almost all biped humanoid robots are controlled by position servo controllers that require joint angle command. This is mainly because good force servo actuators are not commercially available.

To our knowledge, Reference [12] is the first report of a biped robot demonstrating balance against large disturbances. Based on a simple linear inverted pendulum, this controller assumes ZMP fixed to some point and calculates the necessary compensatory joint motions in real-time, then applies a local position controller. One method [13] is based on a similar inverted pendulum model, but it allows ZMP to move according to the state of the COM. These methods also differ in the manner in which the disturbances or COM states are measured: the former measures contact forces directly, whereas the latter measures the ground reaction force. An additional difference is the weight assignment for the compensating DOF: the former uses trunk compensation; the latter uses waist compensation. Mixed weight assignment has also been proposed in Reference [14]. This concept was extended to whole-body motion control in [15] [16]. For example, [15] employs the Jacobian inverse

$$\dot{q} = J_P(q)^\# (-\dot{\overline{r}}_P) \tag{8.57}$$

for some desired COM velocity $-\dot{\overline{r}}_P$. Herein, $J_P^\# = W^{-1} J_P^T (J_P W^{-1} J_P^T)^{-1}$ is the weighted pseudo-inverse with the weighting matrix W, which should be specified in advance. This type of controller works well if the external forces are measured.

Usually, these methods assume GRF measurement; the controller can compute the compensatory motions according to the desired GRF or COM position.

However, if the external force is applied to some point at which the force sensor is not installed, we encounter a problem. For example, we assume that a disturbance is applied to the waist. The joints are position-controlled. Therefore, the disturbance changes the COM as well as the ZMP. The method [13] modifies the desired ZMP to recover the balance. Then, to achieve the actual ZMP track to the desired ZMP, the controller accelerates the waist position. That is, if we push the robot waist, the waist immediately counteracts the external force. This behavior is undesirable for humanoid interaction for safety reasons and differs from what a human does. Although the controller (8.57) can handle weight assignment, we cannot anticipate, in principle, which joint must move in compliance to the unknown external forces.

8.7   ROUGH TERRAIN BALANCING

8.7.1   COMPLIANT TERRAIN NEGOTIATION

This section explains the controller's adaptability to rough terrain including inclined surfaces or steps without measuring the contact forces or terrain shape. The rough terrain addressed here means: (1) local or (2) global changing of the ground surface, or (3) their combination. In general, these cases occur in a time-varying manner. We assume the foot is always located above or on the ground before adaptation.

Let us start with a simple example: a rigid block laid on a step, as shown in Figure 8.20, where x is the COM position of the block measured from the right edge of the step.

There are three equilibrium states in this system. If $x < 0$, the block can stay on the step (Figure 8.20a). If $x = 0$, the block cannot stay on the step (Figure 8.20b). If $x > 0$, the block tumbles around the edge of the step. It then stops at the static

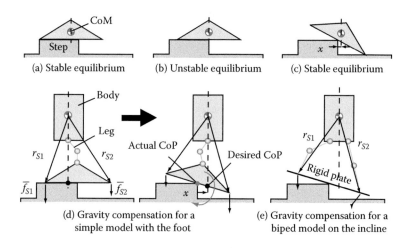

(a) Stable equilibrium   (b) Unstable equilibrium   (c) Stable equilibrium

(d) Gravity compensation for a simple model with the foot

(e) Gravity compensation for a biped model on the incline

FIGURE 8.20   Illustration of ground adaptation mechanism on simple models.

equilibrium, where the block contacts the ground at two points (Figure 8.20c) or tumbles again around the edge of the block itself. The state of the block is sensitive to the position of x, which can vary easily due to the contact forces between the block and the step, or external forces. This is simple physics for demonstrating a natural ground adaptation.

Now, we connect a leg and body (with active joints) to the block and consider the block the robot foot, as shown in Figure 8.20d. Let the position vectors from the COM to the two edges of the foot be r_{S1} and r_{S2}. These are the preset contact points to which the robot applies contact forces f_{S1} and f_{S2}. The ground adaptation mechanism is nicely understood if we consider the gravity compensation case, where the desired COP is set to x and the normal contact forces are linearly distributed onto the contact points based on x, as shown in Figure 8.20d on the left. When $x > 0$, however, the actual COP cannot be positive due to the step, but the desired COP can. This difference generates the moment around the actual COP that results in the foot's rotational motion, as shown in Figure 8.20d on the right. In this way, ground adaptation is based on the COM position, similarly to Figure 8.20a–c. Note that if we shift the desired COP from the COM to the opposite direction, the foot might stay on the step as long as the joints can generate appropriate torques (e.g., when kinematic limits are not exceeded).

Although the stabilization of $x = 0$ in Figure 8.20 is not necessarily the control objective, note what happens if we apply the balancing controller (8.11). Asymptotic stabilization is clearly difficult with this simple balancer. We can only achieve $x = 0$ in the sense of Lyapunov because a rocking motion between (b) and (c) will appear. With an advanced nonlinear controller for an underactuated system, we can persistently achieve $x = 0$ without the rocking motions. The drawback in this case, however, is that highly nonlinear (inverse) dynamics is involved.

The above description also serves to illustrate more complex situations. For example, suppose that the robot is put on a spherical ground surface, as on the left of Figure 8.21, where the contact points are preset to the edge of each foot (four in total). The controller assumes the ground is flat a priori, therefore it generates control torque so that the desired contact points are equally distributed. Then, by the same mechanisms as Figure 8.20d, the foot automatically adapts to the shape of the ground. However, in this case, the foot is flat, whereas the ground is rounded, so only one contact is possible for each foot. To compute the contact equilibrium, detailed analysis "including the ground shape" is required. Note that if the controller only distributes the antigravitational force to the two interior contact points, the feet do not rotate and they do not adapt to the ground shape in this sense.

Let us return to the simple model in Figure 8.20. Now we remove the foot, as in Figure 8.20e. Instead, assume that the robot has two legs with tip positions r_{S1} and r_{S2} and that a mass-less rigid plate is put under the leg. The main difference from the situation in Figure 8.20d is that $|r_{S1} - r_{S2}|$ is allowed to change because there are no constraints between the feet.

Assume active balancer (8.11) is applied to achieve $x = 0$. If the inclination is fixed, then the robot may achieve asymptotical stabilization: $x \to 0$. If the incline is changing, the robot compliantly changes its posture to adapt to the ground. However, in this case it may be difficult to maintain $x = 0$ without a dynamic model of

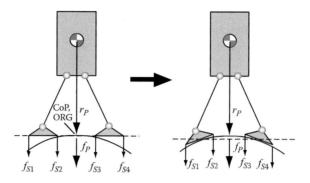

FIGURE 8.21 Adaptation to unknown uneven ground surface (with enough friction). (Left) Initial configuration. (Right) After adaptation. Adaptation reflects feedforward downward forces (antigravitational forces) applied to (preset) contact points.

inclination. It depends solely on how fast the controller can predict the ground shape and how precisely it can compute the required joint torque.

Finally, let us discuss the combined case. For example, we consider Figure 8.22, where a robot model with flat feet is standing on the ground, which is suddenly inclined. We suppose the active balancer (8.11) is always applied so that the COM is the origin, defined as the center of the feet. It immediately computes recovery GAF to pull the COM back to its origin. At the same time, the orientation of each foot is automatically adapted to the inclination by the distributed contact forces, as shown in Figure 8.20d. Therefore, ground adaptation in this case is achieved by a combination of active (feedback) balancing control and our contact force distribution scheme.

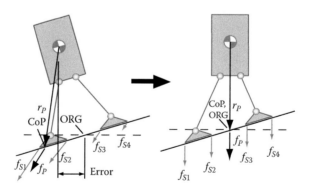

FIGURE 8.22 Adaptation to unknown inclination. Adaptation reflects combination of feedback controller (8.11) that brings the COM back to the origin (ORG) and feedforward ground adaptation (Figure 8.11 + Figure 8.21).

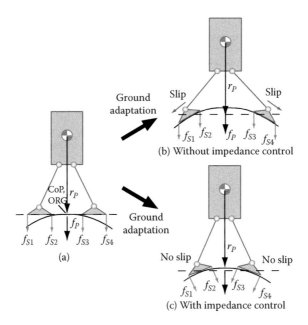

FIGURE 8.23   Adaptation to unknown uneven surface with low friction. (a) Initial configuration. (b) Foot adapts to ground surface, but a slip is caused by low friction. (c) Equilibrium state caused by impedance control.

8.7.2 IMPEDANCE CONTROL FOR LOW FRICTION ADAPTATION

This section considers the case where the ground friction force is not adequate. As described in Section 8.3.4, when $\overline{f}_{xP} = 0$ and $r_{xP} = r_{yP} = 0$, the optimal contact force distribution law yields no horizontal contact force $\overline{f}_{xS}, \overline{f}_{yS}$.[*]

Consider again the case when the robot is put on a rounded surface, as shown in Figure 8.23a. Because the friction force is not sufficient (the ground surface is slippery), the robot may need internal horizontal forces between each foot, especially when the inclination of the surface is large. Clearly, without horizontal internal forces the robot will have difficulty maintaining its posture, as shown in Figure 8.23b.

If the condition of the ground surface, including the levels of the irregularity and friction coefficient, is known in advance, the necessary internal forces can be calculated. However, in general, it is not known. Therefore, we introduce impedance control for friction adaptation by simply generating internal forces in proportion to the distance of the feet.

To avoid defining new symbols, here we give a simple example. Suppose there are two feet (or hands) making contact with the ground. We define the contact index set $I_\alpha = \{1, 2, \dots, 2\bar{\alpha}\}$ and divide it equally into two sets: $I_\alpha = I_{\alpha 1} \cap I_{\alpha 2}$ with

---

[*] Otherwise, there are internal forces between each foot because the optimal contact force distribution, generate uneven horizontal contact forces, which are linearly distributed around the desired COP. See Figure 8.11.

$I_{\alpha 1} = \{1, 2, \ldots, \bar{\alpha}\}$ and $I_{\alpha 2} = \{\bar{\alpha}+1, \bar{\alpha}+2, \ldots, 2\bar{\alpha}\}$.* Considering the X-coordinate only, we set the desired internal force $\bar{\lambda}_{xSj}$ as:

$$
\bar{\lambda}_{xSj} = \begin{cases} -\frac{K_S}{\bar{\alpha}}(x_{Sj} - x_{S(j+\bar{\alpha})} - \bar{x}_{12}), & j \in I_{\alpha 1} \\ -\frac{K_S}{\bar{\alpha}}(x_{Sj} - x_{S(j-\bar{\alpha})} - \bar{x}_{12}), & j \in I_{\alpha 2} \end{cases}
\tag{8.58}
$$

where $\bar{x}_{12} \geq 0$ is the desired step distance between the feet, and K_S is the desired stiffness.

Internal force $\bar{\lambda}_{xS} = [\bar{\lambda}_{xS1}, \bar{\lambda}_{xS2}, \ldots, \bar{\lambda}_{xS\alpha}]$ yield to the (modified) desired contact forces:

$$
\begin{bmatrix} \bar{f}_{xS1} \\ \bar{f}_{xS2} \\ \cdots \\ \bar{f}_{xS\alpha} \end{bmatrix} = A_x^{\#} \begin{bmatrix} \bar{x}_P \\ 1 \end{bmatrix} \bar{f}_{xP} + \begin{bmatrix} \bar{\lambda}_{xS1} \\ \bar{\lambda}_{xS2} \\ \cdots \\ \bar{\lambda}_{xS\alpha} \end{bmatrix}.
\tag{8.59}
$$

Similarly, we compute $\bar{\lambda}_{yS}$.

In this way, the collective internal force forms an impedance around the desired step distances between the feet (hands). The desired step distances are set to a preferred value. For example, one can set the values by the distances of the feet when one of the contacts for each foot has been detected, and this holds for all feet. Note that the "actual" support convex hull is not determined by the desired step distances, but is shifted to where the internal and friction forces are in balance. A possible equilibrium state is illustrated in Figure 8.23c.

Although our method does not require ground information, it is not applicable to a very low friction surface unless K_S has been already set extremely high based on the prediction (possibly by vision information). A larger K_S may cause undesirable resonance during the swinging phase, thus from a practical point of view it may be reasonable to increase or decrease the stiffness based on the sensitivity of the slip to the ground condition. Adding integral action (PI control) has the same effect.

We show two terrain adaptation experiments. The first experiment is on adaptability to uneven ground while maintaining the robot's balance. The result is shown in Figure 8.24. In this example, a wooden block is put under its right foot and suddenly removed or returned by an operator. Sometimes the right foot changes its posture due to on the actual contact forces on the foot. In this particular experiment, the impedance controller (8.58) was deactivated to allow the foot to move freely when it does not make contact with the ground. Figure 8.25 shows the corresponding experimental data. The top graphs show the COM and the desired COP positions. The origin is defined as the center of the feet. Limits on the desired COP are indicated by two dashed lines, ± 0.135 m (half of the foot length). The feedback gains and damping are the same as in Section 8.5.

The left bottom graph of Figure 8.25 is the height of each foot (the origin is the center of both feet), showing the irregularity of the ground, and the right bottom graph shows the attitude of the base (pelvis) measured by a gyro sensor.

---

* For example, $I_{\alpha 1} = \{1, 2, 3, 4\}$ and $I_{\alpha 2} = \{5, 6, 7, 8\}$ for a biped case shown in Figure 8.23b.

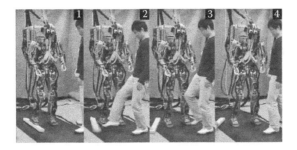

FIGURE 8.24   Balancing on (unknown) uneven ground. A wooden block is put under the right foot and then removed by the operator, who sometimes pushes the robot. The robot maintains its balance by measuring the center of mass and applying recovery contact forces while compliantly changing its posture, showing adaptability to rough terrain. See Figure 8.25.

The second experiment was on adaptability to a time-varying inclination. The result is shown in Figure 8.26. Even though the human operator randomly changes the inclination of the wooden plate, the robot keeps its balance while changing its postural configuration. In this example, the impedance controller (8.58) is used with $K_S = 1,000$ N/m because the friction of the wooden plate is not sufficient. The task feedback gain for (8.11) and joint damping were the same as in the above experiment.

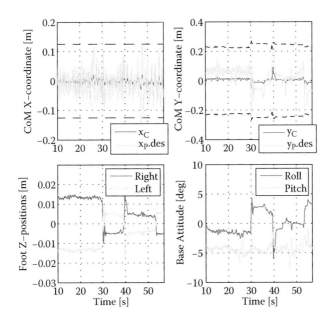

FIGURE 8.25   Experimental data (task-space variables) corresponding to Figure 8.24. (Top) $x_P.des$ and $y_P.des$ indicate the desired COP, synchronized to COM movements x_C and y_C. (Bottom) Sudden changes in foot height and body orientation (Euler angles) show irregularity of ground. Note that the origin of the task-space coordinates is located at the center of the feet.

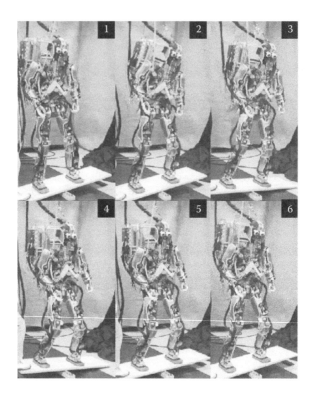

FIGURE 8.26   Balancing on a wooden seesaw. Inclination is manually changed at random by human operator. The robot keeps its balance by measuring the center of mass and applying recovery contact forces while compliantly changing its posture, showing adaptability to rough terrain. Photographs are at every two seconds. See Figure 8.27 for details.

The maximum ground inclination was $\pm 15.64°$ in this experiment. Note that we do not explicitly specify any of the joint angles, which is a common property of task-space force control. As seen from the figure, the COM tracking error in the Y-direction is large. Note that the error is due to the external forces caused by the sudden change of the ground surface. In fact, the error in the X-direction is small compared to the Y-direction.

If the operator inclined the surface quickly, then both feet slipped down the slope and roughly maintained the distance between the feet. When the operator inclined the surface too quickly, one foot lost contact completely, and the robot started to tumble before the balancer compensated for the position disturbance.

There are two main reasons for the errors in the COM: (1) measurement noise and delay, and (2) imperfect joint torque control. Currently, we are using a 3-Hz Butterworth filter for COM velocity measurement. There is also a delay from the gyro sensor that uses a low-pass filter. These cause delays of postural feedback. The main cause for (2) lies in the identification error in the mass distribution and the error in posture measurements. Because these errors affect the Jacobian calculations, they result in contact-force tracking error. Different from (1), these errors can be reduced

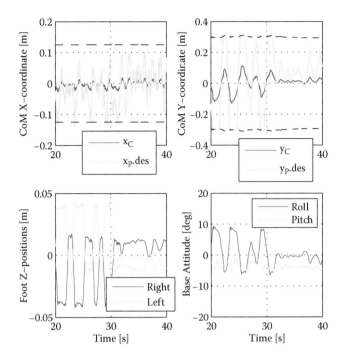

FIGURE 8.27   Experimental data: (Top) x_P.des and y_P.des indicate desired COP, synchronized to COM movements x_C and y_C. Note that y_P.des sometimes saturates at the feet's outer edges (± 0.3 m on average). (Bottom) Changes in foot height and body orientation (Euler angles) indicate inclination of the wooden plate. Maximum ground inclination was $\pm 15.64°$ in this experiment.

by improved calibration or adaptive compensation including integral action. On the other hand, in simulations perfect kinematic measurements are available, hence the tracking errors can be greatly reduced by increasing the PD gains.*

In addition to adaptation experiments, we conducted simple balancing experiments on a very slippery floor to demonstrate the effectiveness of our impedance controller. We put hydraulic oil on the floor. Before the experiment, we confirmed that a human could not walk on this surface. Initially, the robot was standing at an initial posture with high position gains. In this case, the robot was not actively balancing, and one could easily move the robot on the ground because there was little friction between its feet and the ground. Then the robot smoothly switched to force control mode and compliantly balanced. If we applied impedance control in addition to nominal balancing control, the robot could keep its stance while balancing. In contrast, if we removed the impedance controller, then the robot slipped as soon as a small disturbance was applied.

---

*Limitations are caused by the unknown dynamics of the ground contact, simulated by a simple spring-damper model.

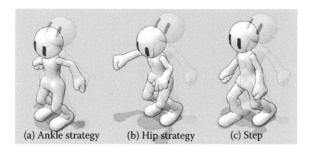

(a) Ankle strategy   (b) Hip strategy   (c) Step

FIGURE 8.28   Three push-recovery strategies when suddenly pushed forward (the initial postures are the same). Each name is conventionally used in the physiology and biomechanics literature.

8.8   PUSH RECOVERY REVISITED

When a human is suddenly pushed forward, he or she can take various actions to recover balance. Physiologists have been discussing the three major recovery strategies: (a) ankle strategy, (b) hip strategy, and (c) step [17] [18]. Because human postural control is fundamental full-body motor control for humans, a high-level nervous system is supposed to be involved as well as vestibular, somatic, and proprioceptive feedback [19–22]. Actually, the ankle strategy is not just joint-stiffness control because there should be antigravitational forces, which can be computed by using an internal model of the body such as mass distribution of the each limb.

The challenge is to suggest computational models of human postural control by exploiting possible controller realizations on compliant humanoid robots, which interact with the environment in real-time as humans do [23]. Such an attempt is also beneficial to development of human-friendly assistive/rehabilitation devices. Our approach first implements the above three strategies on humanoid robot models, and then compares the results with human data at different levels of sensory-motor control.

In robotics, these controllers have been discussed, but in most cases they are independently studied without the integration issue (see the biped walking literature). One of the open questions is how humans combine (or just switch) these three modules according to the sensory feedback and internal model. A model predictive controller is proposed in Reference [24], and some model-based criteria are proposed in [25] [26], based on human model and human gait data.

Our motivation is a unified viewpoint of postural balancing, which covers ankle strategy to hip strategy as follows:

- From the causality of the controlled system, the motion of the center of mass is determined by the ground reaction force, and the control problem is how to distribute the desired GRF into full-body joint torque. Directly solving the full dynamics leads to a rich balancing controller, which we call the dynamic balancer (DB) in this section. This balancer is free from so-called "ZMP (zero moment point) criteria" [10], and can be applied to the

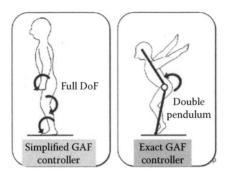

FIGURE 8.29 Overall control architecture of postural balancing. This is composed of dynamic balancer, static balancer, gravity compensation, and joint stiffness. The four blocks on the left show each control module, whose details are shown in the indicated sections and equations (ignore some overlapped terms in the equations). $\bar{r}$ is the desired task-space position, and $\bar{q}$ is the desired joint-space position. The outputs of the modules are simply summed up according to the task objectives.

    underactuated situation, the situation which "hip" strategy implies. However, it suffers from the computational cost, internal motions, and singularity as shown in Section 8.10.

- Within the limitation of the ZMP criteria, we can use a passivity-based postural control strategy without taking the dynamic effect into account. At the cost of reducing the GRF (hence, the balancing ability), the internal motions due to the redundancy can be suppressed by simple joint damping, which also suppresses the angular momentum around the COM. These make the ZMP almost equivalent to the center of pressure. We call this balancing controller the static balancer (SB) in this section. This controller lies between the ankle strategy and hip strategy.

    This section presents a simple integration of SB and DB and applies to a human-sized humanoid robot. Specifically we discuss a superposition of the control outputs (joint torques) of the two control modules (Figure 8.29). Because we have no solid evidence on how humans use these two modules, we first implement and test both on the actual robot, and then get back to the theory. The motivation to use SB is its simplicity, practicability, and safety. However, it loses the controllability of the COM once its projection locates outside SR. On the other hand, DB suffers from difficulties in implementation: it is difficult to derive the dynamic model, and difficult to compute (measure) the necessary dynamics to compensate. Therefore, we use the SB as a default, and combine the DB in a safe way.

    Specifically, we take a simple superposition of SB and DB, where the DB is computed for a reduced two-link acrobot [27] model, where the control input is available only for the hip joint (ankle joint is free). The two-link model is the reduced model of the humanoid robot, where the first link is composed of the shank and thigh, and the second link is the torso.

    The control output from DB is just superposed to the hip joint. Similar integration can be found in Reference [28], but a difference is that we limit the desired COP for SB,

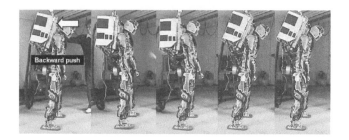

FIGURE 8.30   Typical behavior of superposition of SB and DB.

but do not do that for DB to take advantage of its strong recovery performance. Also, we always use the full-body joint torque with the optimized contact force control.

Figure 8.31 shows an experimental result with the same controller. Strong forward/ backward pushes have been applied from the human operator to the robot pelvis. Graph (e) shows the torque output from the two-link DB module, which is distributed to both right and left hip joints. A large input is required for a "strong" backward push applied at $t = 33$ s (indicated by bPush2). The required torque for the hip joint exceeds 200 Nm. On the other hand, in the case of a similar "strong" forward push at 42.5 s (indicated by fPush3), the hip torque is about −150 Nm. The compensation appears at the knee joint (Graph (c)).

8.8.1 STEPPING

When the stepping motion is triggered, the swinging foot starts to track to a desired trajectory planned online to prevent the robot from falling. The trajectory was given by the 3D symmetric walking control (SWC) law with a fixed stride [29]. Let $r_Q = [x_Q, y_Q, z_Q]^T \in \mathcal{R}^3$ be the position vector from the COM to the center of the swinging foot. SWC constrains the motion of the swinging leg by

$$\begin{bmatrix} x_P + x_Q \\ y_P + y_Q \end{bmatrix} = 0 \tag{8.60}$$

For some given $\overline{d}_{xy}$ (design parameter), the desired height of the swinging foot is given as a smooth curve of the stride $d_{xy} := \sqrt{(x_Q - x_P)^2 + (y_Q - y_P)^2}$. It starts from zero, passing through a desired height, and terminating at zero when $|d_{xy}| \geq \overline{d}_{xy}$. Note that during walking the direction of the heading plane may change, and the robot move its foot to the direction of the heading [30]. Therefore, SWC makes the projection of the COM locate at the center of the supporting foot whenever the foot touches down. The mathematical properties of SWC presented in Reference [30] are met in the same way by considering the angle variables θ_1 and θ_2 in Figure 8.32. That is, the controlled walking gaits are not asymptotically stable, but Lyapunov stable symmetric orbits. Therefore, for every neighborhood of a walking gait, there exists a family of similar gaits with different energy levels.

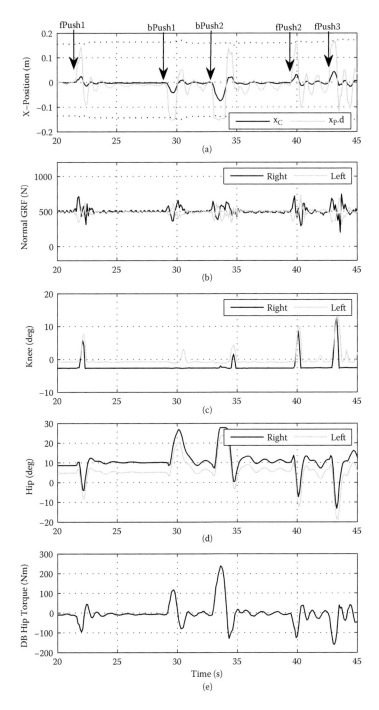

FIGURE 8.31 Experimental data of push-recovery control by SB and DB. Large peaks mean that three forward pushes and two backward pushes (fPush1, bPush1, bPush2, fPush2, fPush3) have been applied to the robot pelvis by the human operator. When bPush2 is applied, the actual hip torque is saturated (not shown in graphs).

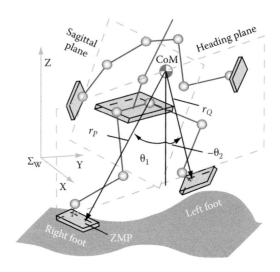

FIGURE 8.32   Illustration of 3D symmetric walking: r_P and r_Q are the vectors from the COM to ZMP and to the center of swinging foot, respectively. θ_1 and θ_2 are the associated angle variables.

For disturbance rejection control, however, in addition to SWC we always apply the balancing controller described in Section 8.10 to achieve asymptotic stability. From the passivity-based control viewpoint, the balancing controller can be regarded as a damping injection [9]. Note that the target COM is always located at the center of the supporting region.

This section shows some selected simulation results. Our control scheme is primitive and the results are premature in the sense that we did not consider the self-collision between the limbs. Therefore, the applicable disturbances are very limited. We use a simplified simulation model without arms and head. The desired COM height is set to 0.85 m, and the horizontal positions are always set to the center of the support. The feedback gains of the superimposed attitude control for the upper body are all 10 Nm/rad.

Figure 8.33 and Figure 8.34 show simulation results of the disturbance rejection with a large disturbance. The robot initially stands at the double support with its right leg put forward. Then a backward push of the magnitude 550 N is applied to the abdomen for 0.2 s. Because the robot tries to keep its balance using the feedback controller (8.11), the desired COP rapidly shifts to the back left edge of the left sole as can be seen from the time profile of $x_P.d$ and $y_P.d$. However, because the external force is too large, the COM x_C rapidly moves backward. As a result, at 2.15 s (0.5 s earlier than the external push is released) the actual ground reaction force of the right leg crosses zero, and the event to stepping is triggered. After the single stepping motion, at 2.75 s the robot again tries to balance at the new supporting state. Finally, the COM converges to the center of the supporting region. Figure 8.35 and Figure 8.36

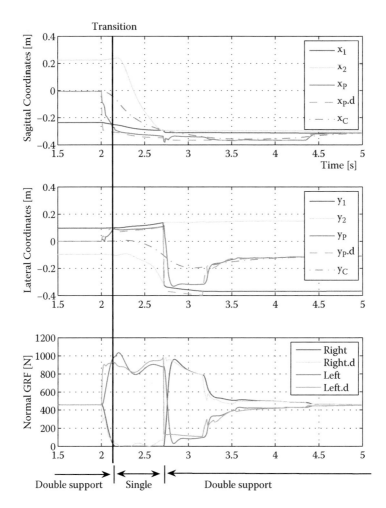

FIGURE 8.33 Disturbance rejection simulation 1. The robot is suddenly pushed backward. The force is of the magnitude 550 N and applied to the abdomen at 2 s (the application time is 0.2 s). The COP (x_P, y_P) are shown with their desired values (indicated by ".d"). The center location of the supporting foot and the swinging foot are indicated by (x_1, y_1) and (x_2, y_2), respectively. (x_C, y_C) represent the COM.

show a similar simulation result, but the magnitude of the disturbance is enlarged to 650 N. This time the robot takes two steps to stop completely.

Note that in both simulations errors can be found between the actual GRF and the desired GRF (=−GAF), hence, between the actual ZMP and the desired one too. The errors come from the uncompensated nonlinear dynamics. At steady state they coincide with each other. Note again that we are using the measured GRF only for the decision making of the transition, but not for the balancing controller.

Figure 8.30 shows one of the experimental results of the push-recovery motion.

FIGURE 8.34 Animation corresponding to Figure 8.33: the robot takes one step to stop. Two red markers show CoM and its ground projection respectively, while the yellow and green ones indicate the desired CoP and the actual one respectively.

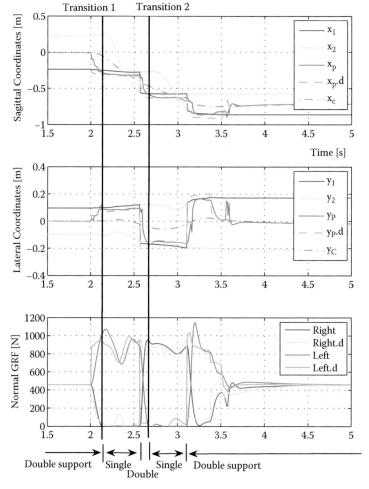

FIGURE 8.35 Disturbance rejection simulation 2. The robot is suddenly pushed from the front by 650 N. See Figure 8.33 for the footnotes.

FIGURE 8.36   Animation corresponding to Figure 8.35: the robot takes two steps to stop.

8.9   QUASI-STATIC WALKING

Without any modification of the controller, one can make the robot transit from double support (DS) to single support (SS) just by shifting the desired COM position. One can use any target COM trajectories moving the center of SR toward the edge of one foot at DS, then change the contact point assignment (see Section 8.4). As for the transition trigger, one can use the event when the COM velocity exceeds the threshold determined by the saddle of the inverted pendulum model. Readers may prefer other conditions depending on their motion task. The robot is gravity compensated, therefore not only the swinging leg, but also the supporting leg has compliance during one-foot balancing.

The transition from SS to DS (touchdown) is done in a feedforward manner; the robot simply switches its controller from SS balancing to DS balancing. At the instance, the swinging foot is assigned with nonzero contact forces, then it starts moving toward the ground even if the desired a trajectory is not given. This is a powerful strategy that compliant robots only can take. Figure 8.38 shows an example. Thanks to the innate terrain adaptability of our controller, the robot can land or even stand on wooden blocks of moderate size.

A combination of balancing on one foot and positioning of the swinging foot leads to static walking. The position of the swinging foot can be commanded by some path planning center, or by interactive devices such as a joypad. The latter is adopted in Figure 8.34. With this simple scheme, climbing stairs and forward/backward walking

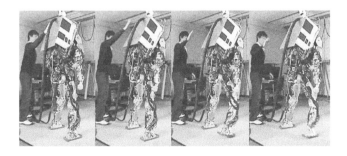

FIGURE 8.37   Combined push-recovery with stepping.

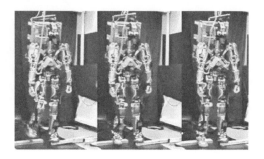

FIGURE 8.38   Transition between SR and DS. The robot is balancing on wooden blocks that are unknown to the robot.

have been achieved without any difficulty. At the beginning of the phase transition from SS to DS, there is an instance when the gCOM lies outside the SR. This is why we call this walking motion quasi-static, whereas in static walking gCOM is always inside SR.

REFERENCES

1. Greenwood, D.T. *Classical Dynamics*. New York: Dover, 1977.
2. Sciavicco, L. and Siciliano, B. *Modeling and Control of Robot Manipulators*, 2nd edition. London: Springer, 1996.
3. Hyon, S., Yokoyama, N., and Emura, T. Back handspring of a multi-link gymnastic robot - reference model approach. *Adv. Robot.* **20**(1):93–113, 2006.
4. Pratt, J., Carff, J., Drakunov, S., and Goswami, A. Capture point: A step toward humanoid push recovery, In *IEEE-RAS International Conference on Humanoid Robots*. Genoa, pp. 200–207, 2006.
5. Murray, R.M., Li, Z., and Sastry, S.S. *A Mathematical Introduction to Robotic Manipulation*. Boca Raton, FL: CRC Press, 1994.
6. Arimoto, S., Sekimoto, M., Hashiguchi, H., and Ozawa, R. Natural resolution of ill-posedness of inverse kinematics for redundant robots: A challenge to Bernstein's degrees-of-freedom problem. *Adv. Robot.* **19**(4):401–434, 2005.
7. ——, Physiologically inspired robot control: A challenge to Bernsteints degrees-of-freedom problem. In *IEEE International Conference on Robotics and Automation*, Barcelona, pp. 4500–4507, April 2005.
8. Takegaki, M. and Arimoto, S. A new feedback method for dynamic control of manipulators, *Trans. ASME, J. Dynam. Syst., Measure. Control* 103:119–125, 1981.
9. van der Schaft, A.J. *L2-Gain and Passivity Techniques in Nonlinear Control*. New York: Springer, 1999.
10. Vukobratović, M. and Borovac, B. Zero-moment point—Thirty five years of its life. *Int. J. Hum. Robot.* **1**(1):157–173, 2004.
11. Featherstone, R. *Robot Dynamics Algorithms*. Boston: Kluwer Academic, 1987.
12. Takanishi, A., Takeya, T., Karaki, H., and Kato, I. A control method for dynamic biped walking under unknown external force. In *IEEE International Workshop on Intelligent Robots and Systems*, Tsuchiura, Japan, July, pp. 795–801, 1990.

13. Hirai, K., Hirose, M., Haikawa, Y., and Takenaka, T. The development of the Honda humanoid robot. In *IEEE International Conference on Robotics and Automation*, Leuven, Belgium, May, pp. 1321–1328, 1998.

14. Yamaguchi, J., Soga, E., Inoue, S., and Takanishi, A. Development of a bipedal humanoid robot-control method of whole body cooperative dynamic biped walking. In *IEEE International Conference on Robotics and Automation*, Detroit, May, pp. 368–374, 1999.

15. Sugihara, T. and Nakamura, Y. Whole-body cooperative balancing of humanoid robot using COG Jacobian. In *IEEE/RSJ International Conference on Intelligent Robots and System*, Vol. 3, Lausanne, Sept., pp. 2575–2580, 2002.

16. Kajita, S., Kanehiro, F., Kaneko, K., Fujiwara, K., Harada, K., Yokoi, K., and Hirukawa, H. Resolved momentum control: Humanoid motion planning based on the linear and angular momentum. In *IEEE/RSJ International Conference on Intelligent Robots and Systems*, Las Vegas, pp. 1644–1650, 2003.

17. Horak, F.B. and Nashner, L.M. Central programming of postural movements: Adaptation to altered support-surface configurations. *J. Neurophysiol.* **55**(6):1369–1381, 1986.

18. Winter, A.D. *A.B.C. (Anatomy, Biomechanics and Control) of Balance During Standing and Walking*. Waterloo Biomechanics, 1995.

19. Squire, L.R., Berg, D., Bloom, F., Lac, S.D., Ghosh, A., and Spitzer, N.C. Eds., *Fundamental Neuroscience*. Orland, FL: Academic Press, 2008.

20. Peterka, R.J. Sensorimotor integration in human postural control. *J. Neurophysiol.* **88**(3):1097–118, 2002.

21. van der Kooij, H., Jacobs, R., Koopman, B., and Grootenboer, H. A multisensory integration model of human stance control. *Biol. Cybern.* **80**(5):299–308, 1999.

22. Morasso, P.G. and Sanguineti, V. Ankle muscle stiffness alone cannot stabilize balance during quiet standing. *J. Neurophysiol.* **88**(4):2157–62, 2002.

23. Kawato, M. From "Understanding the Brain by Creating the Brain" towards manipulative neuroscience. *Philosoph. Trans. Roy. Soc.* **363**(1500):2201–2214, 2008.

24. Morasso, P.G., Baratto, L., Capra, R., and Spada, G. Internal models in the control of posture. *Neural. Netw.* **12**(7–8):1173–1180, 1999.

25. Kuo, A.D. An optimal control model for analyzing human postural balance. *Biomed. Eng. IEEE Trans.* **42**(1):87 –101, 1995.

26. Popovic, M.B., Goswami, A., and Herr, H. Ground reference points in legged locomotion: Definitions, biological trajectories and control implications. *Int. J. Robot. Res.* **24**(12):1013–1032, 2005.

27. Spong, M. The swing up control problem for the acrobot. *IEEE Control Syst. Mag.* **15**(1):49–55, 1995.

28. Stephens, B. Integral control of humanoid balance. In *IEEE/RSJ 2007 International Conference on Intelligent Robots and Systems*, 2007.

29. Hyon, S. and Cheng, G. Passivity-based full-body force control for humanoids and application to dynamic balancing and locomotion. In *IEEE/RSJ International Conference on Intelligent Robots and Systems*, Beijing, Oct. pp. 4915–4922, 2006.

30. Hyon, S. and Emura, T. Symmetric walking control: Invariance and global stability. In *IEEE International Conference on Robotics and Automation*, Barcelona, April, pp. 1455–1462, 2005.

31. Hogan, N. Adaptive control of mechanical impedance by coactivation of antagonist muscles. *Autom. Control. IEEE Trans.* **29**(8):681–690, 1984.

32. Khatib, O. A unified approach for motion and force control of robot manipulators: The operational space formulation. *IEEE J. Robot. Autom.* **RA-3**(1):43–53, 1987.

33. Hyon, S. Compliant terrain adaptation for biped humanoids without measuring ground surface and contact forces. *IEEE Trans. Robot.* **25**(1):171–178, 2009.

34. Hogan, N. Impedance control: An approach to manipulation: Part i - theory. *Trans. ASME, J. Dynam. Syst. Measure. Control.* **107**(1):1–7, 1985.

35. Hyon, S., Osu, R., and Otaka, Y. Integration of multi-level postural balancing on humanoid robots. In *IEEE International Conference on Robotics and Automation*, 1549–1556, 2009.

Section III

Leaping Forward: Toward
Cognitive Humanoid Robots

9 Learning from Examples: Imitation Learning and Emerging Cognition

Yasuo Kuniyoshi

CONTENTS

In contrast to traditional robots, future humanoid robots will not be explicitly programmed to accomplish new tasks in the real world. Instead they will acquire new skills by imitating human behavior. Learning new skills by imitation is a core and fundamental part of human learning, and a great challenge for humanoid robots. This chapter presents mechanisms of imitation learning, which contribute to the emergence of new robot behavior. We first give a detailed description of the general scientific insights underlying the imitation capability of humans. We present a learning model of the mirror neuron system enabling a humanoid robot to imitate first-seen human motions. Then we make further considerations regarding embodiment.

9.1   IMITATION IN DEPTH

For future humanoid robots, the cognitive capability of imitating the behavior of others is essential. Concerning human development, understanding imitation can help in understanding the origins of human intelligence. Imitation can be regarded as a key to higher-order intelligence.

Behavioral imitation is also a central issue in the cognitive and social development of human infants. Questions arise as to what and how to imitate. Exact copying of body motions is not useful because the model and the imitator often do not share either the same body characteristics or the same environmental or task situation. So at the beginning of the whole imitation process, it is necessary to extract meaningful features from the model's behavior, and to reconstruct them by using the imitator's own behavior [8], [9]. This in turn requires shared attention and understanding the concept of similarity among real-world events. In the real world, human behavior and task requirements alter depending on the given situation. For this reason, fixed response patterns are not suitable. Instead, actions adapted to the given situation are needed.

A synthetic study of imitation uses robots to create imitation capabilities. The use of robots for this purpose helps to understand the basics of human intelligence on the one hand, and contributes to realize multifunctional humanoid robots. Imitation and cognitive development are closely interlinked.

Cognitive development is not possible without the interaction between the body and the environment [17], the foundation of embodied cognition or embodiment. This body–environment interaction should not be regarded as a single process because it contains many aspects that depend on each other. In the case of imitation, the interaction dynamics emerges from the following:

- Body schema with sensory–motor mapping
- Shared attention
- Recognition of actions, requiring the extraction of meaningful data

Among these aspects, a body schema and shared attention form the basis for the imitation process. A body schema is a sensory–motor integration about the body. This includes the mapping from proprioception to body posture and motion, to visual, and to tactile stimuli. We realize that this sort of mapping strongly depends on the morphology of the agent, because the morphology sets the (mechanical) constraints on the interaction between the agent's body and the environment, influencing the gathered sensory data.

Shared attention requires the discovery of a common point of interest. Now this point of interest is broadly defined. It can merely be a concrete physical state (e.g., a crack in the wall) or a physical state representing an abstract goal (e.g., a pile of collected leaves representing the task of cleaning up a garden).

Action recognition and its reproduction are partly supported by the mirror neuron system in the human brain.

9.2   LEARNING MODEL OF THE MIRROR NEURON SYSTEM*

9.2.1   ROBOTIC IMITATION SYSTEMS

In the early days of creating robotic imitation systems, researchers focused on specific tasks. For each task, the system had a fixed specific mechanism. The tasks ranged from block manipulation [8], [9], mobile robot navigation [6], [4], over dynamic arm motion in a "kendama" play [15], to head motion [3], [5]. What the imitator was looking at when observing the model's arm gesture was also investigated [13]. In neuroscience, an integrated model of imitation was proposed [1]. Although these previous works show first successful approaches, they do not fully cover the whole cognitive function of imitation. As mentioned already, it is crucial to extract meaningful features and to re-establish them using their own behavior. Because their own behavior is tightly coupled to developmental processes starting from birth, we propose a developmental approach in order to deepen our understanding of the imitation capability. A developmental perspective [2] also helps us to see the imitation capability in the context of other cognitive functions [7]. A theory of development of imitation was established by Piaget [18]. His theory describes the development of human imitation starting from reflexes at a low sensory–motor level and ranging to the symbolic level. His hypothesis of rigid stage transition was problematic and experimentally refuted. However, the idea of bootstrapping interactive behavior by acquiring novel functionality is considered to be important.

9.2.2   OUR APPROACH TO ROBOTIC IMITATION

Inspired by related work and the developmental perspective, we take a synthetic developmental approach. We aim at creating robotic systems that show imitation capability by using bootstrap learning. For this reason, we first have to define appropriate levels of imitation performance. We introduce three important modes of imitation. These modes define what is shared between the model and the imitator. The modes are:

1. Appearance level
2. Action level
3. Purposive task level

On the appearance level, the posture and motions of the body (including a possible extension, i.e., a tool) are shared. The posture and the motions are directly observable. On the action level, the causal relationship between the posture, body motion, and its immediate effect on the target object is shared. On the purposive task level, the overall goal of a certain task is shared. The posture and body motions can be totally different from those of the model as long as they contribute to the achievement of the shared goal.

The modes or levels of imitation reflect themselves in the following example depicted in Figure 9.1. Here, a family* is cleaning a garden. The task is to collect all

* The work presented in this section was originally published as in Reference [12] in 2003.
* We thank Kuniyoshi's family for their cooperation to this research.

FIGURE 9.1 A garden cleaning task serving as example for the levels of imitation.

the fallen leaves. The girl on the very left is copying the posture and motions of her mother on the right. She focuses entirely on the one-to-one copy of the motions. Her actions do not contribute to the task goal because she uses a mop, and her actions with that tool do not alter the positions of the leaves. Her imitative behavior corresponds to the appearance level. The boy in the middle uses a correct tool for sweeping the ground and also copies the motions of the mother. However, he does not take the goal of the task into account, which results in a distribution of the leaves. His imitative behavior corresponds to the action level. The boy on the right next to the mother behaves differently. In contrast to his mother, he is equipped with a different kind of sweeper as well as a dust tray. He also takes a different posture and applies different motions, but his actions are all directed toward the collection of the leaves, a goal which is shared with his mother. Therefore, his behavior corresponds to the purposive task level.

We do not assume that human imitation capability develops rigidly through these three modes. Instead, each of these modes can imply a different set of underlying mechanisms. All of these modes can be active simultaneously and work together seamlessly.

A great challenge would be to develop an artificial system that autonomously acquires all three modes through learning. And there will be a strong connection between the development of imitation and cognitive development.

In our approach, we try to fuse action recognition and generation by using a shared internal representation. Our hypothesis is that the representation is acquired through visually (vaguely) observing self "motor–babbling" during the fetal period. It provides a unified memory of coarse visual–motor–proprioceptive patterns of spontaneous self-bodily movements. After birth this will naturally elicit imitative responses to the movements presented by someone else that look similar to those experienced in the fetal period.

As far as the neonate's brain is concerned, researchers speculate about the existence of an innate mechanism responsible for multimodal representation of body postures. Neonates are already capable of facial and manual imitation, as discovered by Meltzoff and Moore [14]. We hypothesize that the neural representation of the neonate's self–explored sensory–motor learning can be reused in order to imitate a first-seen gestural motion shown by another human. We created a minimalist model

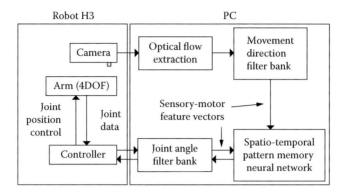

FIGURE 9.2 Overview of our imitation system.

running on a humanoid robot and responding to real human gestures in the real world. Our model reflects the synthetic developmental study of imitation. Moreover, it will clarify whether more complex types of imitation need additional mechanisms. We implemented our model on a real robot described in Section 9.2.3.

9.2.3 SYSTEM DESCRIPTION

9.2.3.1 Overview

In order to provide a synthetic approach to our hypothesis, we conceived a visuo-motor neural system. Our system is depicted in Figure 9.2. It integrates a representation of visual motion as a bank of movement orientation selective filters and a distributed representation of arm joint angles, simulating the corresponding maps in primate brains. In addition, it contains a neural mechanism for high-dimensional temporal sequence learning.

Our system is implemented and runs on a humanoid robot H3 developed by JSK Lab. at the University of Tokyo. The 3-DOF neck carries a CCD camera. The upper body has dual 4-DOF arms with 1-DOF grippers. Our robot is shown in Figure 9.3.

FIGURE 9.3 Our humanoid robot controlled by our imitation system.

9.2.3.2 Representation of Sensory Data

The neonatal brain might contain many cross-modal connections. We hypothesize that a neonatal brain merges visual patterns with proprioceptive and motor patterns, and extracts the information structure in a modal manner. Our system proceeds similarly. It combines the gathered visual and motor feature vectors into a single high-dimensional vector. As we already mentioned, our model contains a neural mechanism for temporal sequence learning. When using such a neural network, it is important to choose a proper encoding scheme of sensory–motor data because this encoding influences the learning ability of the network. The encoded data should preserve the topology of the real data, but on the other hand, a task-specific encoding scheme, such as identification of a chunk of temporal data as a known primitive, should be avoided.

9.2.3.3 Encoding of Visual Features

We selected optical flow to be the basic feature representing visual motion, inasmuch as optical flow plays a central role in motion recognition. The optical flow subsystem extracts flow vectors at 400 points from a 256×256 image at frame rate. We integrated a model of motion direction selective cells to form an optical flow array. The biological version of these cells works in the macaque brain area MT and in the human brain area V5. Overlapping circular receptive fields cover this optical flow array. Each receptive field weights a flow vector with a Gaussian function. A flow histogram for 12 directions is calculated, delivering 12 scalar values for each receptive field. The working principle of our optical flow subsystem is shown in Figure 9.4. The flow vector at the ith point is described by $\mathbf{p_i}$, and the unit vectors in 12 directions are described by $\mathbf{u_j}$ ($1 \leq j \leq 12$). Then the jth directional component of the ith flow f_{ij} is described by:

$$
f_{ij} = \begin{cases} \mathbf{p_i} \cdot \mathbf{u_j} & (\text{if } \mathbf{p_i} \cdot \mathbf{u_j} \geq 0) \\ 0 & (\text{if } \mathbf{p_i} \cdot \mathbf{u_j} < 0) \end{cases}
$$

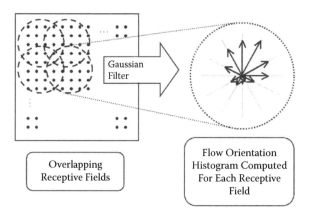

FIGURE 9.4 Motion direction selective cells encode optical flow.

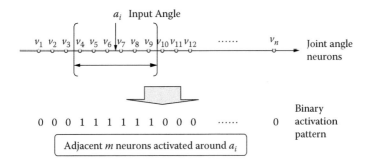

FIGURE 9.5 Binary neurons encode joint angles.

The directional selective cell response is calculated by the Gaussian weighted sum of the above components within the receptive field. The distance of the ith point from the center of the receptive field is denoted by r_i. Then for i ranging within the receptive field, we obtain:

$$F_{kj} = \sum_i f_{ij} e^{-ar_i^2}$$

9.2.3.4 Encoding of Proprioceptive Features

Our system contains a module to extract and encode proprioceptive data. Considering our robot, the joint angles can be described by a four-dimensional vector. A bidirectional interface communicates the joint angles with an arm controller. The encoding of these angles should be robust to noise; that is, small noise in the encoded joint angle should not have much effect on the decoded angle. Of course, the encoding of joint angles has to be reversible, so that a given neural pattern can be decoded into a real joint angle commanded to the robot arm. Therefore, our model uses a distributed encoding scheme. A set of neurons encodes the angle of a DOF. Within that set, the activation of the ith neuron is denoted by g_i. The DOF angle is denoted by a_j. Because we apply a Gaussian distribution for neural activity, the encoding is described by:

$$g_i = ce^{-b(v_i - a_j)^2}$$

where v_i is the ith reference angle (see Figure 9.5), b and c are constants, and e is the exponential function. We consider a binary activation of the encoder neurons, so a constant number of neurons around a_j are active, (see Figure 9.5).

9.2.3.5 Learning of Spatiotemporal Patterns

Our system contains a module for learning and generation of sensory–motor sequences. This module is a nonmonotonic neural network proposed by Morita [16]. This network is a dynamic continuous-valued Hopfield net. It contains neurons with the following nonmonotonic activation function,

$$f(u) = \frac{1 - e^{-cu}}{1 + e^{-cu}} \cdot \frac{1 + \kappa e^{c'(|u|-h)}}{1 + e^{c'(|u|-h)}} \tag{9.1}$$

This activation function is plotted in Figure 9.6.

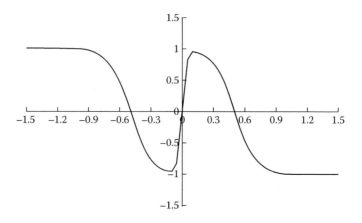

FIGURE 9.6 Nonmonotonic activation function.

The dynamics of the neural potential is described by:

$$\tau \frac{du_i}{dt} = -u_i + \sum_{j=1}^{n} w_{ij} y_j + z_i$$

where u_i denotes the internal potential of the ith neuron, w_{ij} is the connection weight from the jth neuron, z_i is the external input signal, and τ is the time constant. Our network consists of 1,000 neurons, each connected to 499 other neurons. The network uses the covariance rule for learning, which is described by:

$$\tau' \frac{dw_{ij}}{dt} = -w_{ij} + \alpha l_i y_j$$

The learning data are l_i and the learning time constant is τ' with $\tau' \gg \tau$.

But how can one visualize this process of learning spatiotemporal patterns? A visualization of the network behavior helps one gain a deeper insight. Recall that a standard Hopfield network stores a spatial pattern as a point attractor. The operating principle of the nonmonotonic neural network also relies on attractor dynamics. However, the nonmonotonic neural net converges into a trajectory attractor instead of a point attractor. A visualization of the network state space shows that during the learning process, the nonmonotonic neural net creates a trajectory attractor by stringing together many point attractors, each representing a single spatial pattern of the sequence. The creation of trajectory attractors in the network state space is shown in Figure 9.7.

The nonmonotonic activation function (Equation (9.1)) does not let the attractor shape become too steep. So this special activation function contributes to a smooth transition between network states and to an avoidance of fake memories.

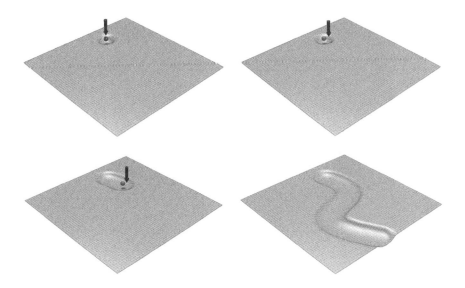

FIGURE 9.7 Formation of a trajectory attractor (From Morita, 1996. With permission.)

9.2.3.6   Recall of Spatiotemporal Patterns

A motion performed in front of the robot is represented by a sequence of visual feature vectors. This lets the network state converge into the previously learned trajectory attractor. The state is trapped into the attractor trench and moves along it. During the motion of the network state along the attractor, the system remaps the current state to the motor vector. The motor vector is decoded into arm joint angles commanded to the robot.

9.2.3.7   Integration of the Neural Network

We found out that we cannot directly attach the nonmonotonic neural network to our sensory–motor subsystems. We conducted experiments with simulated data and checked the association between a spatiotemporal sensory pattern and a corresponding spatiotemporal motor pattern. The network associates correctly up to 30% noise on the sensory pattern. This proves the robustness of Morita's network. However, the performance decreases in case of abrupt data changes or no changes at all. The network should also be helped in learning longer sequences without increasing the number of neurons. For these reasons, we introduced an input module as a kind of preprocessor; see Figure 9.8. Within this input module, FIR filters implement delay units. These delay units smooth out the abrupt changes in the input sequence. The outputs of the delay units are connected to recurrent random remapper neurons. These remapper neurons increase the dimensionality of the input vector. The recurrent connections add inertia to the temporal change in the input pattern. The output of each remapper neuron is connected to one neuron in the nonmonotonic net.

With the help of this input module, we integrated the non-monotonic neural network into our overall system.

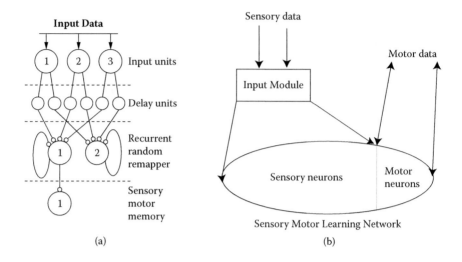

FIGURE 9.8 (a) Input module and (b) integration of the nonmonotonic net.

9.2.4 IMITATION EXPERIMENTS

At the beginning, our robot starts to generate three kinds of arm motions. The motions are:

1. Swinging the hand horizontally
2. Swinging the hand vertically
3. Performing a circular motion

Our robot performs these motions, observes them with its camera, and learns the corresponding sensory–motor sequences. Once the robot has finished learning, it is ready for imitation. A human stands in front of the robot and shows one of the hand motions from the above set. At the same time, our robot observes the human hand motion and directs the arm according to the neural network output. Figure 9.9 summarizes this robot experiment.

We recorded some of the joint angle profiles of the arm motion during the self-learning phase; see Figure 9.10.

In addition, we recorded the profiles of the generated arm motion in response to given human gestures. This was done for three trials for each type of arm motion; see Figure 9.11.

As can be seen in Figure 9.11, our imitation system successfully recognized and regenerated first-seen human arm motions in 7 out of 9 trials. Overall, we conducted 20 trials and we found out that the rate of correct imitation was 81%.

9.2.5 SUMMARY OF THE IMITATION MODEL

Our hypothesis was that the neural representation of the neonate's self-explored sensory–motor learning can be reused in imitating a first-seen gestural motion shown

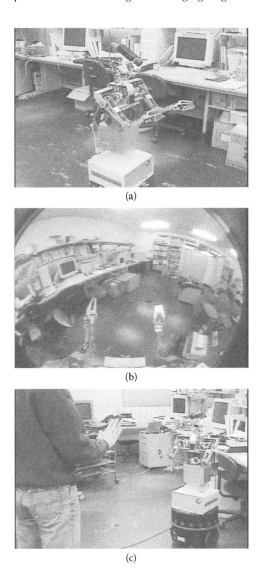

FIGURE 9.9 (a) Our robot performs the arm motions and observes its arms with its camera. (b) The robot's own perspective during the learning is shown. (c) The robot imitates a human's hand gesture.

by another human. We investigated this hypothesis by creating an imitation system on a humanoid robot. Our robot starts to produce self motion patterns. During the exploratory self-learning phase, the robot learns coherent visual–motor patterns by representing them as high-dimensional trajectory attractors. After the exploratory self-learning phase, a human comes in front of the robot. The human performs an arm motion that looks similar to one of the motions executed during the learning process. This visual pattern activates the corresponding visual–motor trajectory attractor, that

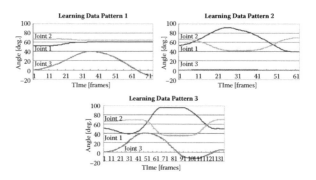

FIGURE 9.10 Pattern 1 corresponds to the horizontal swing, pattern 2 to the vertical swing, and pattern 3 to the circular motion.

is, the robot's memory. This immediately evokes an arm motion represented by the trajectory attractor, resulting in an imitative response. We can say that our robot has acquired the capability of early imitation by self-learning. An important thing to note is that the robot does not recognize what it sees as a human hand, nor implement any specific computation dedicated to imitation. When the robot is imitating the human motion, it is just re-experiencing the visual–motor pattern that it has learned before about self-motion. This outcome substantiates our above hypothesis.

9.3 EMBODIED BASIS OF IMITATION

The recognition of actions of others is supported by the mirror neuron system in primates' brains. In the previous section, we showed how a simple sensory–motor learning about self-motor babbling can acquire the basic functionality of the mirror neuron system. However, the system in the previous section assumes that initial motion primitives are already built-in. This is a serious problem because what the system learns is totally determined by the set of motor primitives that are arbitrarily chosen and programmed by the designer of the system. For the system to have an open-ended imitation capability, it is crucial that the system can autonomously discover appropriate action units which can be stably reproduced by another agent. In this sense, the action units constitute a symbolic information structure that can be shared by different agents. It can be communicated to other agents. The shared information (aka "knacks") robustly tells different agents how to achieve the same action on slightly different embodiments. Moreover, they provide goal-related information as they are critical conditions for reaching the goal state. This bridges the mirror neuron level of action units and the goal-directed organization of multiple actions. For our concept of action recognition and reconstruction by using such symbolic embodied information structures, we conducted a robotic experiment. We chose a roll–and-rise action (Figure 9.12), asked subjects to perform this action repeatedly, and measured their whole-body motion with an optical motion capture system.

We analyzed the results in terms of state-space trajectories. These state-space trajectories converged or branched at some critical states. These states can be shared

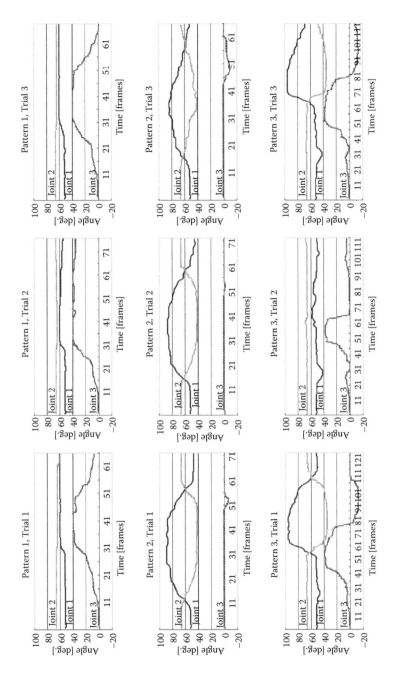

FIGURE 9.11 Recorded joint angle profiles of the imitation result. These are the generated arm motions in response to the three arm motions provided by a human. The first row shows three trials for a horizontal swing. The second row shows three trials for a vertical swing. The last row shows three trials for the circular motion. All these trials were successful except for pattern 1, trial 2 as well as pattern 3, trial 2, where the system failed to recognize the given motions in the middle of the sequences.

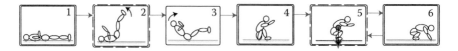

FIGURE 9.12 The roll–and–rise motion as an example for action recognition and reconstruction.

by different agents and facilitate reproducing the action. We can regard these critical states as symbols. We developed a humanoid robot (Figure 9.13) with highly anthropomorphic embodiment. Its outer shape, joint configuration, and mass distribution are as close as possible to an adult human. Recall that morphology is important because it influences the dynamics of the generated action. We tested the action reconstruction on this robot. The roll–and–rise action was already represented by symbols encoding the previously described critical states within the state-space trajectory. Numerical simulation and search generated motor patterns, which conformed to the symbols. We applied these generated motor patterns on the robot, which successfully reproduced the roll-and-rise action; see Figure 9.14.

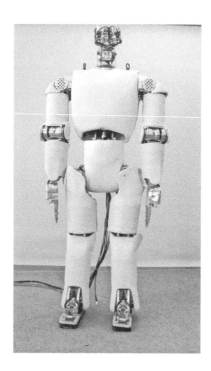

FIGURE 9.13 Our humanoid robot developed for humanoid action execution.

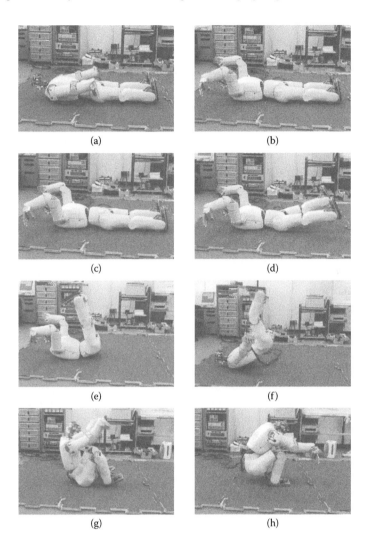

FIGURE 9.14 Our humanoid robot successfully reconstructs the roll-and-rise action.

9.4 CHAOTIC EXPLORATION OF EMBODIED INFORMATION STRUCTURES

Sensory–motor mapping such as visual–motor mapping is an agent's key internal mechanism influencing the agent's behavior. A proper sensory–motor mapping serves as a control mechanism to drive the agent's body. When the agent moves, information structures can emerge, which are used for both control and recognition. A problem is how to control the body when the emergent information structure is not known. A classic method for this problem assumes a given set of *motion primitives*. These are basic actions that can be combined to form different motion sequences leading to arbitrary behavior. A drawback is that the number of possible combinations increases dramatically for many DOFs to control as is the case with humanoid robots. Another drawback is related to the optimization for only one evaluation function neglecting other possibilities. Therefore, we propose a new model. This model does not work with any primitives or evaluation functions. Instead, it makes use of the principle of embodiment. A robot driven by this model exhibits motion patterns consistent with the dynamics of its body. The motion pattern is stabilized by mutual entrainment. In addition, the model is adaptive because it can switch to other motion patterns in the case of changing environmental constraints. This model consists of a distributed set of chaotic elements coupled with the multiple-element musculo-skeletal system [11].

9.5   CONCLUSION

We showed that imitation capability is a complex cognitive function. It requires the extraction of meaningful features and their re-establishment them by using their own behavior. Inspired by the neonate's early imitation capability, we implemented a learning model of the mirror neuron system on the robot. It acquires early imitation capability based on self-explored sensory–motor learning. Our model acquires an internal representation of the visual–motor features extracted during self-motions. The internal representation serves for both perception and action generation of simple hand gestures. Then we pointed out that the units of actions that the learner explores should not be externally given. Rather they should be grounded on the embodiment of the learner itself. And we showed how meaningful units of actions emerge from humanoid embodiment. The units are "symbolic" in the sense that they are reproducible between different agents sharing a similar embodiment. Finally we briefly presented how the learner can start with no predefined primitives, explore, and discover such action units emerging from embodiment by means of embodiment as a chaos coupling field. Integration of these principles would lead to an experimental setup with realistic humanoid embodiment, chaotic exploration, and sensory–motor learning. Currently we are progressing in this research direction with a simulation of a human fetus in utero with a self-organizing nervous system model [10].

REFERENCES

1. Arbib, M.A., Billard, A., Iacoboni, M., and Oztop, E. Synthetic brain imaging: Grasping, mirror neurons and imitation. *Neural Netw.* **13**:975–997, 2000.
2. Asada, M., MacDorman, K.F., Ishiguro, H., and Kuniyoshi, Y. Cognitive developmental robotics as a new paradigm for the design of humanoid robots. *Robot. Auton. Syst.* **37**(2–3):185–193, 2001.
3. Berthouze, L., Bakker, P., and Kuniyoshi, Y. Development of oculo–motor control through robotic imitation. In *IEEE/RSJ International Conference on Robotics and Intelligent Systems, Osaka*, pp. 376–381, 1996.
4. Dautenhahn, K. Getting to know each other — Artificial social intelligence for autonomous robots. *Robot. Auton. Syst.*, **16**:333–356, 1995.
5. Demiris, J., Rougeaux, S., Hayes, G.M., Berthouze, L., and Kuniyoshi, Y. Deferred imitation of human head movements by an active stereo vision head. In *Proceedings of*

the IEEE Int' l Workshop on Robot and Human Communication (ROMAN), pp. 88–93, 1997.

6. Hayes, G.M. and Demiris, J. A robot controller using learning by imitation. In Borkowski, A. and Crowley, J.L., Editors. *Proceedings of the 2nd International Symposium on Intelligent Robotic Systems*, Grenoble, pp. 198–204, LIFTA-IMAG, 1994.

7. Kuniyoshi, Y. The science of imitation — Towards physically and socially grounded intelligence —. Tech. Rep. TR-94001, RWC, 1994.

8. Kuniyoshi, Y., Inaba, M., and Inoue, H. Learning by watching: Extracting reusable task knowledge from visual observation of human performance. *IEEE Trans. Robot. Autom.* **10**(5), 1994.

9. Kuniyoshi, Y. and Inoue, H. Qualitative recognition of ongoing human action sequences. In *Proceedings of IJCAI93*, pp. 1600–1609, 1993.

10. Kuniyoshi, Y. and Sangawa, S. Early motor development from partially ordered neural–body dynamics — Experiments with a cortico–spinal–musculo–skeletal model. *Biol. Cybern.* **95**(6):589–605, Dec. 2006.

11. Kuniyoshi, Y., Suzuki, S., Sangawa, S. Emergence, exploration and learning of embodied behavior. In *Thrun, S., Brooks, R., Durrant-whyte, H. (Eds.), Robotics Research: Results of the 12th International Symposium Isrr.*, Vol. 28 of Springer Tracts in Advanced Robotics, New York: Springer Verlag, 2007.

12. Kuniyoshi, Y., Yorozu, Y., Inaba, M., and Inoue, H. From visuo–motor self learning to early imitation — A neural architecture for humanoid learning. In *Proceedings of the IEEE International Conference on Robotics and Automation*, pp. 3132–3139, 2003.

13. Mataric, M.J. and Pomplun, M. Fixation behavior in observation and imitation of human movement. *Cogn. Brain Res.* **7**(2):191–202, 1998.

14. Meltzoff, A.N. and Moore, M.K. Imitation of facial and manual gestures by human neonates. *Science*, **198**:75–78, 1977.

15. Miyamoto, H., Schaal, S., Gandolfo, F., Gomi, H., Koike, Y., Osu, R., Nakano, E., Wada, Y., and Kawato, M. A kendama learning robot based on bi–directional theory. *Neural Netw.* **9**(8):1281–1302, 1996.

16. Morita., M. Memory and learning of sequential patterns by nonmonotone neural networks. *Neural Netw.* **9**:1477–1489, 1996.

17. Pfeifer, R., Scheier, C. Understanding Intelligence. *Cambridge, MA: MIT Press*, 1999.

18. Piaget, J. Play, Dreams and Imitation in Childhood. New York: W. W. Norton, 1962.

10 Toward Language: Vocalization by Cognitive Developmental Robotics

Minoru Asada

CONTENTS

One of the big issues of human cognitive development is the language acquisition process, and constructive approaches have been attacking the issue to understand this process by building vocal robots. This chapter gives a survey on these approaches. First, cognitive developmental robotics is briefly introduced to show a general idea of these approaches. Then, learning of vowel vocalization by parrotlike teaching is shown as one of the models of the interaction between an infant and a caregiver. Next, the caregiver's anticipation during the interaction process is analyzed, and as a result, it is shown how the caregiver shapes the infant vowels. Finally, future issues are discussed toward language acquisition by cognitive developmental robotics.

10.1 MEANING OF CONSTRUCTIVIST APPROACHES

The main purpose of the constructivist approach is to generate a completely new understanding through cycles of hypothesis and verification, targeting the issues that are very hard or almost impossible to solve under existing scientific paradigms. A typical example is evolutionary computation that virtually recreates a past we cannot observe, and shows the evolutionary process (e.g., [8]). If we reduce the timescale, the ontogenetic process, that is, the individual development process, can be the next target for constructivist approaches. Development of neuromechanisms in the brain or cognitive functions in infants are at considerably different levels. The former have their own history as developmental biology and the researchers' approach to the issue under this discipline. The latter deal with cognitive development in developmental psychology, cognitive science, and so on. Depending on the age, given (already acquired) functions, and faculties to be acquired through interactions with the environment, including other agents, should be clearly discriminated.

The subjectivity argued by Hashimoto et al. [8] is only meaningful if we focus on human individuals. That is, the process of self-establishment by infants provides various kinds of research issues in developmental psychology, cognitive science, and sociology, including the issue of communication. Therefore, approaches to the issues that are difficult to solve under a single existing discipline might be able to be found by the constructivist ones. Especially, for the infant's cognitive development, developmental psychology largely depends on observations from outside (macroscopic), neuroscience tends to be microscopic, and brain imaging is more difficult to apply to infants than adults. Thus, a single paradigm seems difficult to approach, and is it not easy to verify hypothesized models. Then, it is time for the constructivist approaches to begin to play an active role.

10.2   COGNITIVE DEVELOPMENTAL ROBOTICS AS A CONSTRUCTIVIST APPROACH

A representative constructivist approach is CDR (its survey is given by Asada et al. [1]). References [26] and [29] take similar approaches, but CDR puts more emphasis on human/humanoid cognitive development. A slightly different approach is taken by Atkeson et al. [2] who aim to program humanoid behavior through the observation and understanding of human behavior and by doing so, give a clearer insight into the nature of the latter. Although partially sharing the purpose of human understanding, they do not exactly deal with the developmental aspect.

Figure 10.1 summarizes the various aspects of the development according to the survey by Lungarella et al. [18] from the viewpoints of external observation, internal structure, its infrastructure, and social structure, focusing especially on the underlying mechanisms in different forms.

Roughly speaking, the developmental process consists of two phases: individual development at an early stage and social development through interaction between individuals at a later stage. The former relates mainly to neuroscience (internal mechanism), and the latter to cognitive science and developmental psychology (behavior observation). Intrinsically, both should be seamless, but there is a big difference between them at the representation level for the research target to be understood. CDR aims not at simply filling the gap between them but, more challengingly, at building a new paradigm that provides new understanding of ourselves while at the same time adding a new design theory of humanoids that are symbiotic with us. So far, CDR has been mainly focusing on computational models of cognitive development, but in order to understand more deeply how humans develop, robots can be used as reliable reproduction tools in certain situations such as psychological experiments. The following is a summary:

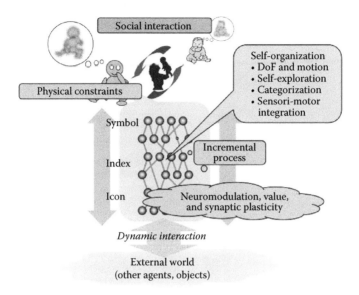

FIGURE 10.1 Various aspects of the development from viewpoints of external observation, internal structure, its infrastructure, and social structure. (Based on Lungarella et al., 2008. With permission.)

A: Construction of computational model of cognitive development

1. *Hypothesis generation*: Proposal of a computational model or hypothesis based on knowledge from existing disciplines.
2. *Computer simulation*: Simulation of the process difficult to implement with real robots such as physical body growth.
3. *Hypothesis verification*: With real agents (humans, animals, and robots), then go to 1.

B: Offer new means or data to better understand human developmental process → mutual feedback with A.

1. Measurement of brain activity by imaging methods.
2. Verification using human subjects or animals.
3. Providing the robot as a reliable reproduction tool in (psychological) experiments.

According to the two approaches above, there are many studies inspired by the observations of developmental psychology and by evidence or findings in neuroscience. The survey by Asada et al. [1] introduces these studies based on the constructive model of development they hypothesize.

Based on the idea of CDR, the author's group started the JST ERATO Asada Synergistic Intelligence System Project (www.jeap.jp) in 2005 as a total 6.5-year project. A small part of the achievements is summarized in Figure 10.2 and also in Reference [1]. In this project, a variety of robot platforms has been developed and utilized for the studies of different aspects of human cognitive development.

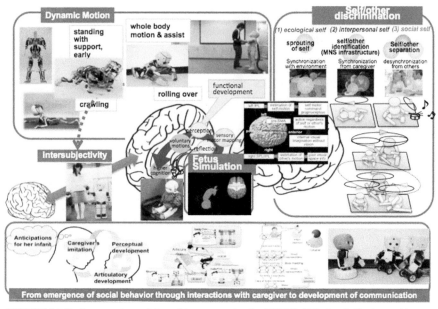

FIGURE 10.2 An overview of the achievements by the JST ERATO Asada Project with some new projects. See color insert.

Figure 10.3 shows these robots. From the left to the right, CB^2 (child robot with biomimetic body: 1.3 m, 33 kg, 56 DoFs, air-cylinder–type actuators, vision, auditory, and tactile sensors), Neony (top: .5 m, 3.5 kg, 22 DoFs, 90 tactile sensors), Synchy (bottom: .3 m, 2.3 kg, LED, 17 DoFs), Kindy (1.1 m, 27 kg, 42 DoFs), Noby (9-month old baby: .75 m, 9 kg, 30 DoFs), and Pneuborn 7 and 13 (PMA (pneumatic muscle actuator) robots: 7: .8 m, 4 kg, 30 DoFs).

One of the social interactions we have been attacking is vowel imitation toward language acquisition with an aspirational vocalization mechanism based on the idea of physical embodiment. In the following sections, an artificial vocal system with learning methods through the interaction with a caregiver and the analysis of the caregiver's bias are introduced and discussed toward language acquisition.

FIGURE 10.3 Platforms developed by the JST ERATO Asada Project.

10.3 DEVELOPMENT OF VOCAL IMITATION

It has been reported that children of eight months can imitate an adult's single vowel [12]. To reach such a developmental milestone of imitation, infants should not only acquire sensorimotor mapping to vocalize sound but also find the correspondence of utterances between themselves and their caregivers. It has been a central interest of developmental science how infants acquire the abilities underlying these requirements. Infants' ability of listening to adult voices appears in a language-independent manner from birth and gradually adapts to their mother tongue [30]. Meanwhile, infants' utterances are first quasi-vocalic sounds that resemble vowels and gradually adapt to their caregiver's ones [16] along with descent of the epiglottis [27]. Therefore, it seems likely that vocal interaction with their caregivers is needed for such an infant to adapt its vocal system to the caregivers' language. However, how and what kinds of interaction among infants' learning mechanisms and the caregiver's behavior are essential for the processes are still unclear.

On the other hand, recent imaging technology has started locating early sensitivities for language input in the infant brain [4,5]. However, it remains difficult to investigate the links among these sensitivities and caregiver's interaction through the developmental course because of the limitations of current imaging technology. Similar difficulties also exist in other approaches based on observation due to ethical problems to control infant development, therefore synthetic approaches are expected to contribute to find the missing links.

Some synthetic studies have been conducted to model what happens in the vocal babbling period that has often been considered to have an important role in speech acquisition. Guenther et al. [7] have developed a neural network model called the DIVA model that learns sensorimotor mapping through random exploration resembling infants' babbling, and argued its correspondences to results of adult fMRI work. Westermann and Miranda [31] have developed another model that incorporates physical embodiment in sensorimotor learning so that it can acquire reliable vowel articulations through a babbling period while being exposed to ambient language. Kanda et al. [13] have proposed a recurrent neural network model that automatically segments continuous sounds of vowel sequences to learn sensorimotor mapping and pointed out the importance of incorporating self-articulation in the segmentation process. Hörnstein and Santos-Victor [9] have considered multiple phases through the babbling period by regarding the caregiver's behavior during that time and argued the same importance on recognition of others' vowels. However, the above synthetic studies have not paid much attention to the caregiver's behavior which has started from birth and seems essential in vocal development.

Kokkinaki and Kugiumutzakis have reported an important characteristic of the caregiver's behavior for infants to learn such correspondences: parents imitate their infants at a high rate in the first six months [14]. As implied from other observations where imitation of an infant's utterances by caregivers is induced by the infant's vowellike utterances [19] and inversely encourages such utterances [25], such parental imitation or being imitated might play an important role in the developmental process of vocal imitation. The importance of being imitated has been demonstrated in synthetic studies of computer-simulated vocal agents although they did not directly aim at modeling infants' development. It has been shown that a population of learning agents with a vocal tract and cochlea can self-organize shared vowels among the population through mutual imitation [3,24]. However, these previous works have assumed that the infant model could produce the same sound as the caregivers' one if it learned proper parameters of articulation unlike the situation of infants due to their immaturity. In other words, they have paid less attention to another developmental hurdle of finding the correspondence of utterances between themselves and their caregivers.

On the other hand, Yoshikawa et al. [32] have addressed this issue in human–robot vocal interaction and demonstrated the importance of being imitated by the human caregiver, whose body is different from the robot's, as well as subjective criteria of the robot such as ease of articulation. With a similar experimental setting, Miura et al. [20] have argued that being imitated by a caregiver has two meanings: not only informing of its correspondence to the caregiver but also guiding the robot to performing it in a more similar way to how the caregiver performs.

Inspired by the previous work [20], Ishihara et al. [10] have computationally modeled an imitation mechanism as a Gaussian mixture network (GMN), parts of which parameters are used to represent the caregiver's sensorimotor biases such as the perceptual magnet effect [15], which is adopted in the computer simulation of Oudeyer to evolve a population to share vowels [23]. The perceptual magnet effect indicates a psychological phenomenon where a person recognizes a stimulus as a more typical one of closer categories that the person possesses in his or her mind. They have conducted a computer simulation where an infant and the caregiver imitate each other with their own GMNs, where one for the infant is learnable and the other for the caregiver involves a certain level of magnet effects. They have found that the caregiver's imitation with magnet effects could guide infants' vowel categories toward corresponding ones. Interestingly, the effectiveness of guidance was enhanced if there was what they call automirroring bias in the caregiver's perception so that she perceives the infant's voice as closer to the one that resembled her precedent utterance.

Although previous synthetic studies focusing on finding correspondence through being imitated have assumed that the infant is almost always or always imitated by the caregiver for simplicity, it is apparently unrealistic. In such more realistic situations, infants should become able to realize that they are being imitated. Miura et al. [21] have addressed this issue by considering a lower rate of being imitated in computer simulation and proposed a method called autoregulation, that is, active selection of action and data with underdeveloped classifiers of the caregiver's imitation of an infant's utterances.

In many previous synthetic studies, the problems of learning a sensorimotor map and finding correspondence have always been coped with separately and under the situation where only vocal exchanges are assumed. However, to model the developmental process more faithfully, we should consider both problems under a more realistic situation of vocal interaction such as sharing attention or the naming game.

Table 10.1 shows key developmental aspects for vocalization from the viewpoints of neuroscience, developmental psychology, and constructive approach. In the following sections, the details of our approaches [10,21,32] are given.

TABLE 10.1
Key Issues to Develop Vocalization[a]

Neuroscientific Evidence	Psychological Evidence	Constructive Approaches
Descent of the epiglottis:	8 m infants imitate: Jones [12]	DIVA model: Guenther et al. [7]
Sasaki et al. [27]	Adapt to mother tongue:	Physical embodiment:
Locating early sensitivities for	Werker and Tees [30]	Westermann and Miranda [31]
Language input:	Perceptual magnet:	RNN: Kanda et al. [13]
Dehaene–Lambertz et al. [4]	Kuhl and Meltzoff [16]	Population: deBoer [3], Oudeyer [24]
Gervain et al. [5]	Mutual imitation:	Different body: Yoshikawa et al. [32]
	Kokkinaki and Kugiumutzakis [14],	Unconscious guidance: Miura
	Masataka [19], Pelaez-Nogureas [25]	et al. [20]
		Two biases: Ishihara et al. [10]
		Detection of being imitated:
		Miura et al [21]

[a] Extracted from Table IV in Asada et al. 2009.

10.4   VOWEL ACQUISITION THROUGH MOTHER–INFANT INTERACTION

Inspired by the observation that infants acquire phonemes common to adults without having the capability to articulate, nor having prior knowledge about the relationship between the sensorimotor system and phonemes, we took a constructivist approach to building a robot that reproduces a similar developmental process [32]. Two general issues are addressed: what are the interactive mechanisms involved and what should be the behavior of the caregiver/teacher? Based on findings in developmental psychology, it is conjectured that: (a) the caregiver's vocalization in response to infants' cooing reinforces the infant's articulation along the caregiver's phonemic categories; and (b) the caregiver's repetition with adult phonemes helps to specify the correspondence between cooing and the caregiver's phonemes as well as determining the acoustic properties of the phonemes. The robot consists of an artificial articulatory system with a five-degrees-of-freedom mechanical system deforming a silicon vocal tract connected to an artificial larynx, an extractor of formant, and a learning mechanism with self-organizing auditory and articulatory layers. Starting off with random vocalizations, the system uses the caregiver's repetitive utterances to bootstrap its learning. Figure 10.4 shows a vocal robot system that consists of an articulation part and an auditory one with their corresponding layers, respectively. Both layers are self-organized and connected by Hebbian learning through parrotlike teaching by a caregiver. Figure 10.5 shows formant distributions of the robot and the caregiver that indicate how they are different from each other phonemically.

Employing the initially random articulation mechanism causes invariant pairs of units to activate in both layers simultaneously because the caregiver is engaged in repetitive utterances. Therefore, through the learning process, clusters of articulation vectors are matched with corresponding vowels as connections between both layers. However, interactions may connect multiple articulation units with a corresponding

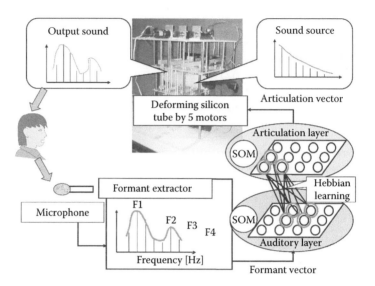

FIGURE 10.4 An overview of the system.

vowel inasmuch as the caregiver may interpret some vocalizations caused by different articulations as the same vowel. To match a heard vowel with a unique articulation in order to vocalize it, we introduced subjective criteria into the learning rule so that the articulation involving less torque and less deformation is selected; that is, the articulation vectors with less torque and less deformation obtain a stronger connection from the auditory layer and vice versa. Therefore, the learning rule for the connections is slightly modified as follows.

Let w_{ij} be a connecting weight between the ith unit in the auditory layer and jth unit in the articulation one. The change of the weight in the normal Hebbian rule is specified as

$$\Delta w_{ij} \propto \alpha_i^f \alpha_j^m \tag{10.1}$$

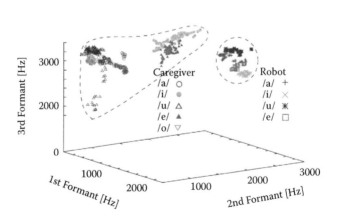

FIGURE 10.5 Formant distributions of the robot and the caregiver.

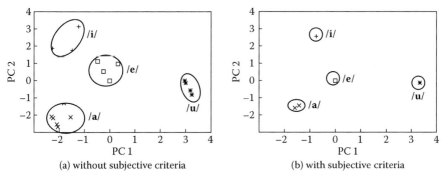

(a) without subjective criteria

(b) with subjective criteria

FIGURE 10.6 Distribution of the articulation vectors in the two major principal component spaces without (a) and with (b) subjective criteria.

where α_i^f and α_j^m denote the activations of the ith and jth units in the auditory and articulation layers, respectively. In order to resolve the arbitrariness in determining proper articulations, the torque to deform the tract and its resultant deformation are minimized. Then, the weight change is redefined as

$$\Delta w_{ij} \propto \eta(c_{trq}, c_{idc})\alpha_i^f \alpha_j^m \qquad (10.2)$$

where $\eta(c_{trq}, c_{idc})$ is a function that evaluates the toil involved in the articulation, and c_{trq} and c_{idc} are the cost functions of the torque to deform the tract and its resultant deformation, respectively.

Figures 10.6a and b show two distributions of the articulation vectors in the two major principal component spaces without and with subjective criteria, respectively. In Figure 10.6a, the articulation vectors corresponding to each vowel are more widely spread out whereas those in Figure 10.6b are well converged into a small region owing to the proposed subjective criteria. Figure 10.7 shows the resultant vocal tract shapes obtained by learning methods (SOM and Hebbian) through the parrotlike teaching of the caregiver.

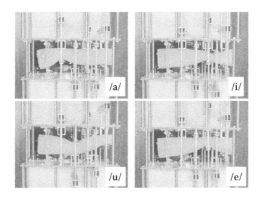

FIGURE 10.7 Acquired shapes of the vocal tract.

We have proposed a learning model of vowel acquisition implemented by a robot with different articulation parameters but without prior knowledge about the relationship between the sensorimotor system and phonemes. Here, we argue several issues in the model. One big question to the approach is why we are not using a speaker instead of such a mechanical vocal system. The main reason is that we would like to realize the physical embodiment as close to us as possible so that we can focus on the issue of cognitive development. Consequently, the subjective criteria designed by physical constraints successfully worked to obtain the converged vowels.

Even if infants can reproduce the utterances of the caregiver as they are, infants perceive their own reproduced soundwaves differently from the caregiver's original one, because these two soundwaves travel different pathways: the caregiver's soundwave travels only through the air and the infant's travels through both the air and his or her body. As a result, these two soundwaves at the infant's auditory sensors are different from each other. This fact makes imitation based on the similarity of raw soundwaves difficult. The current robot is not designed to address this issue, and building a new robot with a microphone inside its body is one of our future works.

In the model, the robot–caregiver interaction is simplified: the caregiver always utters the vowel that matches the cooing of the robot if the cooing can be regarded as a vowel. However, the proposed learning method does not always require the caregiver's repetition because it extracts clusters utilizing the statistical consistency in the data. That is, the method works only if the caregiver tends to be engaged in the repetitive utterances. Furthermore, this simplification is unrealistic because the human caregivers usually talk to infants with adult language, that is, words or sentences. This is another future issue.

The third issue is the behavior of the caregiver who has an important role to lead the developmental process of infant articulation. The next section deals with this issue, that is, how the caregiver's anticipation shapes the infant's vowel through mutual imitation.

10.5   HOW CAREGIVER'S ANTICIPATION SHAPES INFANT'S VOWEL THROUGH MUTUAL IMITATION

The mechanism of infant vowel development is a fundamental issue of human cognitive development that includes perceptual and behavioral development. We modeled the mechanism of imitation underlying caregiver–infant interaction by focusing on potential roles of the caregiver's imitation in guiding infant vowel development [10]. The proposed imitation mechanism is constructed with two kinds of the caregiver's possible biases in mind. The first is what we call "sensorimotor magnets," in which a caregiver perceives and imitates infant vocalizations as more prototypical ones as mother-tongue vowels. The second is based on what we call "automirroring bias," by which the heard vowel is much closer to the expected vowel because of the anticipation being imitated. Computer simulation results of caregiver–infant interaction show the sensorimotor magnets help form small clusters and the automirroring bias shapes these clusters to become clearer vowels in association with the sensorimotor magnets.

To focus on the caregiver's behavior, we assume the following:

1. Iteration of multiple imitative turn-takings
2. Initial categorization of caregiver vowels
3. Statistical learning capability of contingent relation
4. Formant extractor (same as the previous work)
5. Established articulation skills
6. Caregiver's consistent imitation with sensorimotor magnets
7. Caregiver's automirroring bias
8. Infant's unexpressed biases

Suppose that two people alternately iterate and imitate each other's voices (assumption 1), and that the sound can be denoted by an N_s-dimensional vector and the articulation to produce the imitation sound can be denoted by an N_a-dimensional vector.

Figure 10.8 illustrates the imitation process by the proposed mechanism: at the tth step of mutual imitation, it listens to the other's voice $s(t) \in \Re^{N_s}$ and imitates $s(t)$ by articulation $a(t) \in \Re^{N_a}$. This imitation process consists of three functions: an automirroring bias module (b) that biases input sounds, a sensorimotor map module (f) that produces an imitation utterance from biased input, and an anticipation module (g) that calculates what we call "automirroring anticipation" from one's last imitation utterance. "Automirroring anticipation" is defined as the perceptual bias by which other's voices are heard as if they resemble the listener's own precedent utterances because of the listener's anticipation being imitated.

Other's voice $s(t)$ is biased to automirroring anticipation $s^g(t-1)$ and converted to $s^b(t)$ that is given by:

$$s^b(t) = b(s(t), s^g(t-1); \alpha) \tag{10.3}$$
$$= s(t) + \alpha(s^g(t-1) - s(t)) \qquad (0.0 \le \alpha \le 1.0) \tag{10.4}$$

where α is a parameter that determines the strength of the automirroring bias (assumption 7). When α is close to 0, output $s^b(t)$ nearly equals the original

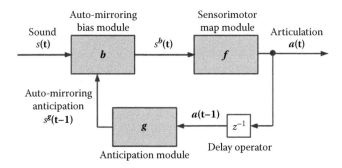

FIGURE 10.8 Proposed imitation mechanism considering biasing elements. (From H. Ishihara et al. *IEEE International Conference on Development and Learning*, 2011. With permission.)

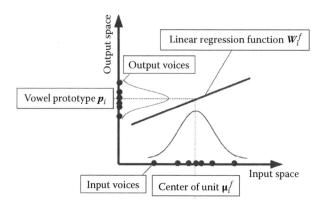

FIGURE 10.9 Illustration of sensorimotor magnets with linear regression function.

input $s(t)$ because the automirroring bias is weak. Conversely, when α is close to 1, output $s^b(t)$ is almost attracted to automirroring anticipation $s^g(t-1)$.

Because human adults and infants (and robots also) do not have completely identical sensorimotor systems, they cannot perfectly reproduce the other's voices. Therefore, these other voices need to be converted into articulation parameters that generate the listener's own utterable vowels. We use the normalized Gaussian network (NGnet) to map the other's utterable vowel region onto the listener's own utterable articulation parameter space (assumption 6). NGnet is a modular probabilistic regression function that maps N_s-dimensional input space onto N_a-dimensional output space with units. Figure 10.9 shows how sensorimotor magnets are illustrated where we suppose that input data are normally distributed with a central focus on the center of an NGnet unit. The distribution of output data belonging to the ith unit is determined by the linear regression matrix W_i^f of the NGnet f.

An anticipation module converts articulation $a(t-1)$ to automirroring anticipation $s^g(t-1)$. We use NGnet g to map N_a-dimensional input space onto N_s-dimensional output space contrary to NGnet f in the sensorimotor map module. Automirroring anticipation is calculated by

$$s^g(t-1) = g(a(t-1); \theta^g) \tag{10.5}$$

where θ^g is a set of parameters of NGnet g.

We assume that a simulated infant (hereinafter infant) initially has an immature imitation mechanism; the parameters of NGnet f (i.e., θ^f) in the sensorimotor map module are estimated through mutual imitation. Before the learning, in other words, her vowel prototypes p_i are not clear vowels by which she cannot accurately imitate utterances of a simulated caregiver (hereinafter caregiver). Furthermore, we assume that she does not have automirroring bias (i.e., $\alpha = 0$) for the simplicity of the first simulation trial (assumption 8). Here the infant's task is tuning parameters to match vowel prototypes with the clearest vowels for a caregiver by mutual imitation.

In the simulations, an infant and a caregiver alternately imitate each other with their imitation mechanisms (assumption 1). The infant has an immature imitation

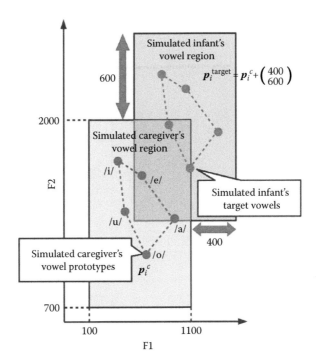

FIGURE 10.10 Settings of two vowel regions of simulated infant and caregiver.

mechanism and updates parameters θ^f of NGnet $\boldsymbol{g}$ with the learning method, and the caregiver has a mature imitation mechanism, that is, her imitation parameters are fixed during a session of iterating mutual imitations (assumption 6).

For the current simulation, the vowel regions both of the caregiver and the infant are determined in two-dimensional vowel space, as shown in Figure 10.10 (assumption 4), so the differences between the caregivers and the infants are highlighted. We also regard this vowel space as the articulation parameter space; that is, the vowels and articulation parameters to generate these vowels are the same two-dimensional vector (assumption 5).

Figure 10.11 shows the differences of the learning results under several conditions where the strengths of the caregiver's biasing elements are different and each distribution is an example of the result under each condition. We simulated interaction under the following conditions:

(a) Where a caregiver has both automirroring bias and sensorimotor magnets
(b) Where a caregiver only has automirroring bias
(c) Where a caregiver only has sensorimotor magnets
(d) Where a caregiver has no biasing elements

In these distributions, red (blue) dots represent the infant voices $\boldsymbol{y}(t)$ (the caregiver voices $\boldsymbol{x}(t)$) in the vowel space in the final 1,000 steps. The apexes of the red (blue)

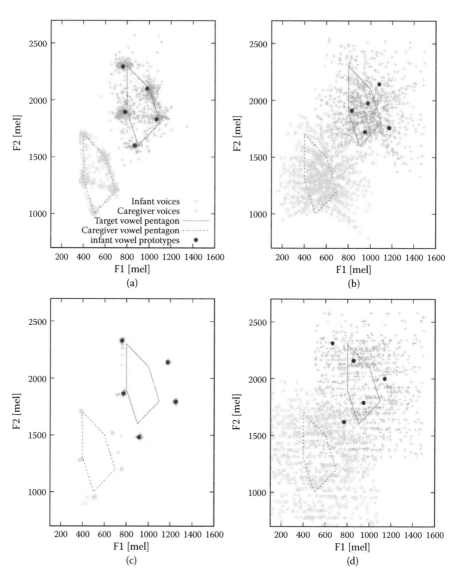

FIGURE 10.11 Difference of learning results under several conditions. Apexes of solid pentagons represent target vowels of infant, in other words, clearest vowels in her vowel region, and black dots represent infant vowel prototypes after learning.

pentagons represent the target vowels of the infant p^{target} (caregiver vowel prototypes p_i^C). Black dots represent the vowel prototypes of the infant after learning. These distributions indicate that the caregiver's biasing elements heavily affected the results of the infant's learning. In (a), the infant vowels are almost correctly converged, whereas (b)–(d) do not seem to be. In (b), the sensorimotor magnet is missing, therefore not convergence but rather divergence of infant vowels and the caregiver's imitated voices is observed. On the other hand, in (c), automirroring biases are missing, and as a result, there is fast convergence but three of five vowels converged onto wrong locations. In the case of no biases, nothing happened, as shown in (d).

We simulated many combinations of the automirroring bias (α_c) and the sensorimotor magnet (β_c), and obtained the best result in the case of the well-balanced combination: that is, $\alpha_c = .45$ and $\beta_c = .55$. Another couple of combinations around it are also fine. This suggests that the emergence of the guidance requires a balanced association between automirroring bias and sensorimotor magnets.

Simulation results indicate that these biasing elements of the caregiver guide the infant vowel prototypes to become clear vowels; the sensorimotor magnets help form small vowel clusters, and the automirroring bias shapes these clusters to become clearer vowels in association with the sensorimotor magnets. The results might imply general importance of the caregiver's anticipation of the infant's ability on guiding various social developments of infants.

In our model, infants need to be imitated for learning correspondence of utterances between themselves and caregivers. However, real caregivers do not always imitate their infants. We have to extend our model so that we can also explain how caregivers' nonimitative feedback affects infant vocal development. Furthermore, we fixed the strength of the automirroring bias of the caregiver during interactions and assumed that an infant does not have automirroring bias. In the next section, we partially touch on these issues.

10.6 VOWEL ACQUISITION BASED ON LEARNER'S AUTOMIRRORING BIAS WITH LESS FREQUENT IMITATIVE CAREGIVER

In the previous section, we supposed that the caregiver always imitates the infant's utterance. However, in real interactions, the caregivers imitate their infants' utterances much less frequently. According to the observation data analyzed by Gros-Louis et al. [6], in the caregivers' responses to their infants from 7 to 10 months, they imitate their infants' utterance about 20% only, and the rest were nonimitative ones. This 20% imitation includes not complete but partial imitation, and also imitation of nonverbal utterance. How can infants acquire their vowels from the interaction with their caregivers who less frequently imitate? Here, we propose an interaction model that finds the correspondence to the caregiver's vowels in such an environment under the following assumptions:

1. As already pointed out, the utterable regions of the caregiver and infant are different, and therefore the infant cannot reproduce the caregiver's voice, but can perceive both the caregiver's and its own formants of the utterances (localize them in the formant space). This enables the infants to learn the correspondences of its own vowels to the caregiver's ones. In the model, the learner represents the corresponding caregiver's vowel candidates as normal distributions in the formant space. In Figure 10.12, the learner's ith vowel candidate $\boldsymbol{u}^{/i/}$ has a link to a corresponding normal distribution of which

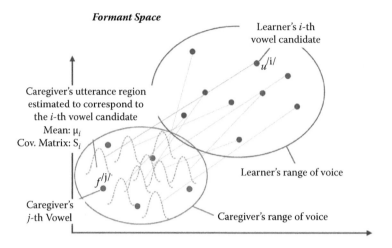

FIGURE 10.12 The relationship between the caregiver's vowels and the learner's primitives.

the mean and covariance matrix are μ_i and S_i that is expected to converge to the caregiver's jth vowel $f^{/j/}$.

2. The learner has m_v vowel candidates, but does not know that the caregiver has M_v ($<m_v$) vowels ($M_v = 5$ for Japanese). The reason why $m_v > M_v$ is that the learner should find M_v vowels corresponding to the caregiver's M_v vowels among m_v candidates.

3. The learner selects two candidates from m_v ones and utters them as continuous vowels inasmuch as a single vowel utterance is not realistic.

4. When the caregiver imitates the learner's utterance, his imitation success probability P_i^S changes according to the clarity of the learner's utterance that leads to the easiness of his imitation. The more (less) clear the learner's utterance is, the higher (lower) is the success probability of the caregiver's imitation. The decision of imitation success or failure is made in terms of each uttered vowel candidate.

5. The caregiver's responses including imitative and nonimitative ones are classified into three types such as, "What did you say?" (no imitation), "Do you like /eo/?" (failed imitation), and "Did you say /au/?" (successful imitation).

Figure 10.13 shows these three cases in the above assumption 5. As stated, the success probability p_i^S depends on the clarity of the vowel candidate uttered by the learner. The imitation probability $p_i = .2$ considering the real situations [6]. A key idea in the proposed model is to apply the automirroring bias to the learner instead of the caregiver as in the previous section where this bias worked to perceive the learner's utterance entrained to the caregiver's vowel by anticipating that the learner always imitates the caregiver's utterance. Here, the learner does not know the vowel correspondence between the caregiver's and its own. Therefore, we do not expect such entrainment. Instead, the learner believes that the caregiver's responses are always an imitation of its own utterance. Key aspects of the automirroring bias on the learner's

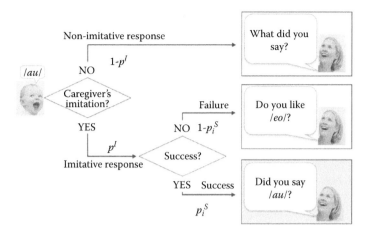

FIGURE 10.13 A flow of probabilistic choice of the caregiver's response.

side are summarized as follows:

- Enhance the corresponding relationship between the caregiver's vowels and the learner's by believing that the caregiver always imitates the learner.
- Enhance the corresponding relationship if the caregiver's response matches better to one of the candidates the learner has than other ones.
- Increase the selection probability of the candidate that is clearer than other ones because its clarity promotes the success of the caregiver's imitation.

The learning is expected to proceed as follows:

1. Distribute m_v normal distributions in the caregiver's voice region, each of which corresponds to one of the vowel candidates of the learner. The value of the distribution indicates the likelihood of how close the caregiver's utterance is to this candidate.
2. Update each distribution with new data based on the caregiver's responses and their imitation likelihoods.
3. Estimate the ease of the caregiver's imitation using the clarity, and select the candidates based on the estimated ease.
4. Go back to 2 unless the number of interactions exceeds the prespecified constant.

In the experiment, we prepared 100 candidates ($m_v = 100$) such that 20 candidates were assigned to one of the caregiver's 5 vowels ($M_v = 5$), and each candidate had a different imitation success probability p_i^S ($p_i^S = 1.0, .95, .90, \ldots, .05$). As the caregiver's responses, we prepared real utterances such as, "What did you say?" "Do you like /au/?" and so on. Therefore, nonimitative responses may have not-uniform distribution of not-imitated vowels, but biased ones which makes the learner work harder to find the correct correspondence. Figure 10.14a and b show the trajectories of

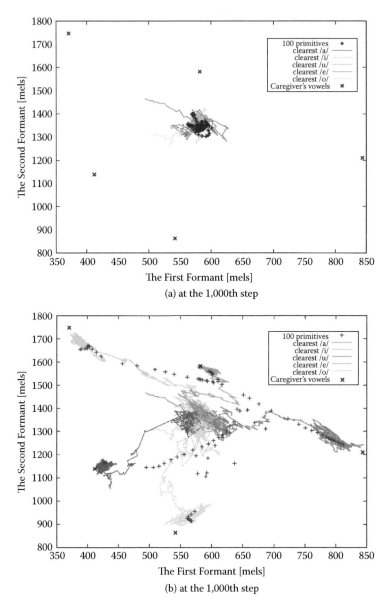

FIGURE 10.14 The trajectories of the centers of Gaussian functions $\mu_1, \ldots, \mu_5$ and values of values of $\mu_1, \ldots, \mu_{100}$ under the condition with the imitation probability $p^I = 0.2$.

the centers of Gaussian functions $\mu_1, \ldots, \mu_5$ (the clearest five candidates) and values of $\mu_1, \ldots, \mu_{100}$ (all candidates) under the condition with the imitation probability $p^I = 0.2$ at the 1,000th and 10,000th steps, respectively. Five crosses indicate the caregiver's vowels, and, "+"s indicate the values of 100 candidates that are initially randomly distributed, but due to the lack of experience (not sufficient data, yet), at the early stage (the 1,000th step (a)), they gathered at the center of the region. However, they gradually extended to the small regions near the caregiver's vowels where many clearer candidates gathered.

Kuhl et al. [17] reported that younger than 6-month-old infants can discriminate all kinds of vowels in any language, but gradually their perception is tuned to their mother's language, and therefore they are losing their perceptual capability before 6 months. Here, we prepared many (20 times) vowel candidates on the learner's side and their normal distributions in the caregiver's utterance regions in order to focus on this losing process from a viewpoint of cognitive developmental robotics. As shown in Figure 10.14b, the resolution of the learner's vowel perception to the caregiver's utterances became high (the centers of the normal distributions gather around the caregiver's vowels) which implies the consequence of the losing process in Reference [17].

In order to evaluate the importance of the learner's anticipation of the caregiver's imitation, we implemented the experiments with different anticipation (strong, normal (used thus far), weak, and very weak. See Reference [21] for more details about the strength definition). Figure 10.15 shows the time course of the average discrepancy between the estimation of the corresponding vowels to the clearest primitive of each category under the condition with the imitation probability $p^I = 0.2$ and different anticipation of the caregiver's imitation. The vertical axis represents a kind of error (the discrepancy), therefore smaller is better. As we can see, the strength of the anticipation of the caregiver's imitation should not be so strong, not be too weak, but be moderate. This seems the same as the automirroring bias on the caregiver

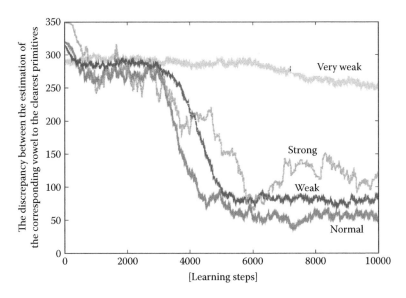

FIGURE 10.15 The transitions of the average discrepancy between the estimation of the corresponding vowels to the clearest primitive of each category under the condition with the imitation probability $p^I = 0.2$ and different anticipation of the caregiver's imitation.

side as mentioned in the previous section. In the case of "very weak," the failure of finding the correspondence at the early stage seriously affects later stage learning. On the other hand, in the case of "strong," the later stage learning is affected. These imply that the anticipation is needed, but after the learner acquired the decision capability for the caregiver's imitation, too much anticipation disturbs finding the correct correspondence. An adaptive scheduling to change the anticipation strength seems needed.

We prepared 100 candidates in the learner's utterance area so that we could simulate the learner's capability of utterance, and make the learner find the correspondence to the caregiver's vowels. However, we have not coped with the development of the learner's utterance capability (extension of the utterance area). We might be able to handle this issue to some extent by replacing the discrete vowel candidates with a probability density function that can be updated (developed) by the anticipation of the caregiver's imitation. This is our future issue.

10.7   TOWARD LANGUAGE ACQUISITION

As a very early stage of language acquisition based on physical embodiment (aspirational vocalization), three studies on vowel acquisition through interactions with caregivers are introduced from a viewpoint of cognitive developmental robotics. They are parrotlike teaching, automirroring bias on the caregiver side, and another automirroring bias on the learner's side.

There are several issues at different levels to be attacked. The first is to improve or integrate three systems together. Automirroring bias only on the caregiver's side is considered whereas the bias on the learner's side is ignored, and vice versa regardless that both exist in the same interaction. This bias should be considered on both sides together, and also be adaptive to the changes caused by learning and development.

In order to focus on vowel vocalization, we restricted to only the auditory channel and articulation in these three studies. However, real interactions include more modalities of sensors and actions. Vision, touch, and musculoskeletal body are typical types of embodiment to be added. We have partially attacked the lexicon acquisition problem by multimodal simultaneous mapping with vocal imitation ([28]), where a caregiver and a learner play a word game by calling object names shown to them and/or imitating these calls to each other, and multimodal mappings between them are learned. However, input data are already categorized symbols including consonants that are another big challenge to pronounce for our vocal robot.

Intrinsic motivation and social interaction are other aspects for an agent to learn the lexicon. Supposing embedded curiosity as one type of intrinsic motivation, active lexicon acquisition is proposed [22], where the degree of curiosity that the robot feels for the objects is determined by the two kinds of saliency: habituation saliency and knowledge. The former saliency is related to the habituation and the latter is related to the strength of the association between the visual features and the labels. A robot changes its attention and learning rate based on curiosity.

In order to examine characteristics and factors of the caregiver's automirroring bias, for example, what kinds of infants' behavior induce this bias in the caregiver's

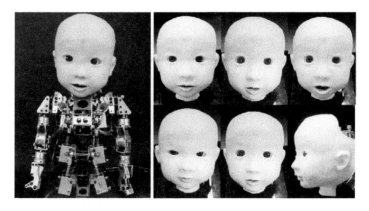

FIGURE 10.16 Affetto head with the internal mechanical structure of the upper torso (left), and examples of the facial expressions (right).

imitation, we need to conduct some psychological experiments. We consider that automirroring bias plays important roles not only in guiding infant vowel prototypes to become clear vowels, but also in maintaining caregiver–infant interaction. We expect that automirroring bias forms an intrapersonal positive feedback loop between the observation to be imitated and the feeling that the opponent is imitating. In their spiral response-cascade hypothesis [32], Yoshikawa et al. suggest the existence of inter– and intrapersonal positive feedback loops, not only between observation and feeling, but also between feeling and action. They explain the mechanism responsible for the emergence and maintenance of communication between agents, not just between a caregiver and an infant.

Considering these issues of mother–infant interaction based on a physical embodiment, we need a much more realistic infant robot to derive emotional states from the caregiver. Ishihara et al. [11] developed such an infant robot, Affetto, to study how the caregiver–child attachment heavily affects the children's developmental pathway. Figure 10.16 shows its appearance, whose head can generate various facial expressions. Currently, Affetto does not have any vocal system, but in future it will have a unified platform.

ACKNOWLEDGMENTS

The author would like to thank the following people for their support of the achievements introduced in this chapter. Professors Koh Hosoda (Osaka University), Yasuo Kuniyoshi (University of Tokyo), Hiroshi Ishiguro (Osaka University), and Toshiro Inui (Kyoto University), Doctors Yuichiro Yoshikawa (Osaka University), Masaki Ogino (Ksanai University), Katsushi Miura (Currently, Fujitsu), and PhD candidates Hisashi Ishihara and Yuki Sasamoto (Osaka University).

REFERENCES

1. Asada, M., Hosoda, K., Kuniyoshi, Y., Ishiguro, H., Inui, T., Yoshikawa, Y., Ogino, M., and Yoshida, C. Cognitive developmental robotics: A survey. *IEEE Trans. Auton. Mental Devel.*, **1**(1):12–34, 2009.

2. Atkeson, C.G., Hale, J.G., Pollick, F., Riley, M., Kotosaka, S., Schaal, S., Shibata, T., Tevatia, G., Ude, A., Vijayakumar, S., and Kawato, M. Using humanoid robots to study human behavior. *IEEE Intell. Syst.*, **15**(4, July/August):46–56, 2000.

3. de Boer, B. Self organization in vowel systems. *J. Phonet.*, **28**(4):441–465, 2000.

4. Dehaene-Lambertz, G., Dehaene, S., and Hertz-Pannier, L. Functional neuroimaging of speech perception in infants. *Science*, **298**:2013–2015, 2002.

5. Gervain, J., Macagno, F., Cogoi, S., Penä, M., and Mehler, J. The neonate brain detects speech structure. *Proc. Nat. Acad. Sci. USA*, **105**:14222–14227, 2008.

6. Gros-Louis, J., West, M.J., Goldstein, M.H., and King, A.P. Mothers provide differential feedback to infants' prelinguistic sounds. *Int. J. Behav. Devel.*, **30**(6):509–516, 2006.

7. Guenther, F.H., Ghosh, S.S., and Tourville, J.A. Neural modeling and imaging of the cortical interactions underlying syllable production. *Brain Lang.* **96**:280–301, 2006.

8. Hashimoto, T., Sato, T., Nakatsuka, M., and Fujimoto, M. Evolutionary constructive approach for studying dynamic complex systems. In G. Petrone and G. Cammarata (Eds.), *Recent Advances in Modelling and Simulation*, Chapter 7. I-Tech Books, 2008.

9. Hörnstein, J. and Santos-Victor, J. A unified approach to speech production and recognition based on articulatory motor representations. In *Proceedings of 2007 IEEE/RSJ International Conference on Intelligent Robots and Systems*, pp. 3442–3447, 2007.

10. Ishihara, H., Yoshikawa, Y., Miura, K., and Asada, M. How caregiver's anticipation shapes infant's vowel through mutual imitation. *IEEE Trans. Auton. Mental Devel.*, **1**(4):217–225, 2009.

11. Ishihara, H., Yoshikawa, Y., and Asada, M. Realistic child robot "Affetto" for understanding the caregiver-child attachment relationship that guides the child development. In *IEEE International Conference on Development and Learning, and Epigenetic Robotics (ICDL-EpiRob 2011)*, CD–ROM, 2011.

12. Jones, S.S. Imitation in infancy - The development of mimicry. *Psychol. Sci.* **18**(7):593–599, 2007.

13. Kanda, H., Ogata, T., Komatani, K., and Okuno, H.G. Segmenting acoustic signal with articulatory movement using recurrent neural network for phoneme acquisition. In *Proceedings of 2008 IEEE/RSJ International Conference on Intelligent Robots and Systems*, pp. 1712–1717, 2008.

14. Kokkinaki, T. and Kugiumutzakis, G. Basic aspects of vocal imitation in infant-parent interaction during the first 6 months. *J. Reproduct. Infant Psychol.* **18**:173–187, 2000.

15. Kuhl, P.K. Human adults and human infants show a "perceptual magnet effect" for the prototypes of speech categories, monkeys do not. *Percept. & Psychophys.*, **50**:93–107, 1991.

16. Kuhl, P.K. and Meltzoff, A.N. Infant vocalizations in response to speech: Vocal imitation and developmental change. *J. Acoustic Soc. Amer.* **100**:2415–2438, 1996.

17. Kuhl, P.K., Williams, K.A., Lacerda, F., Stevens, K.N., and Lindblom, B. Linguistic experience alters phonetic perception in infants by 6 months of age. *Science* **255**:606–608, 1992.

18. Lungarella, M., Metta, G., Pfeifer, R., and Sandini, G. Developmental robotics: a survey. *Connect. Sci.*, **15**(4):151–190, 2003.

19. Masataka, N. and Bloom, K. Accoustic properties that determine adult's preference for 3-month-old infant vocalization. *Infant Behav. Devel.* **17**:461–464, 1994.

20. Miura, K., Yoshikawa, Y., and Asada, M. Unconscious anchoring in maternal imitation that helps finding the correspondence of caregiver's vowel categories. *Adv. Robot.*, **21**:1583–1600, 2007.

21. Miura, K., Yoshikawa, Y., and Asada, M. Vowel acquisition based on an auto-mirroring bias with a less imitative caregiver. *Adv. Robot.*, **26**:23–44, 2012.

22. Ogino, M., Kikuchi, M., and Asada, M. Active lexicon acquisition based on curiosity. In *The 5th International Conference on Development and Learning (ICDL'06)*, 2006.

23. Oudeyer, P.-Y. Phonemic coding might result from sensory-motor coupling dynamics. In *Proceedings of the 7th International Conference on Simulation of Adaptive Behavior (SAB02)*, pp. 406–416, 2002.

24. Oudeyer, P.Y. The self-organization of speech sounds. *J. Theor. Bio.* **233**(3):435–449, 2005.

25. Pélaez-Nogueras, M., Gewirtz, J.L., and Markham, M.M. Infant vocalizations are conditioned both by maternal imitation and motherese speech. *Infant Behav. Devel.* **19**:670, 1996.

26. Sandini, G., Metta, G., and Vernon, D. Robotcub: An open framework for research in embodied cognition. In *Proceeding of the 4th IEEE/RAS International Conference on Humanoid Robots*, p. 13–32, 2004.

27. Sasaki, Levine, Laitman, and Crelin. Postnatal developmental descent of the epiglottis in man. *Arch. Otolaryngol.* **103**:169–171, 1977.

28. Sasamoto, Y., Yoshikawa, Y., and Asada, M. Mutually constrained multimodal mapping for simultaneous development: Modeling vocal imitation and lexicon acquisition. In *The 9th International Conference on Development and Learning (ICDL'10)*, CD–ROM, 2010.

29. Weng, J., McClelland, J., Pentland, A., Sporns, O., Stockman, I., Sur, M., and Thelen, E. Autonomous mental development by robots and animals. *Science*, **291**:599–600, 2001.

30. Werker, J.F. and Tees, R.C. Cross-language speech perception: Evidence for perceptual reorganization during the first year of life. *Infant Behav. Devel.* **25**:121–133, 2002.

31. Westermann, G. and Miranda, E.R. A new model of sensorimotor coupling in the development of speech. *Brain Lang.* **89**:393–400, 2004.

32. Yoshikawa, Y., Asada, M., Hosoda, K., and Koga, J. A constructivist approach to infants' vowel acquisition through mother-infant interaction. *Connect. Sci.*, **15**(4):245–258, 2003.

Index

Note: Page numbers ending in "f" refer to figures. Page numbers ending in "t" refer to tables.